MAKING SPATIAL DECISIONS
USING GIS
AND REMOTE SENSING
A WORKBOOK

Kathryn Keranen ▪ Robert Kolvoord

Esri Press Academic
REDLANDS | CALIFORNIA

To my husband, children, and grandchildren—you provide a constant source of joy to me.

—KK

To the Geospatial Semester students and teachers for making classrooms a spatial place.

—RK

CONTENTS

PREFACE

MAKING
SPATIAL
DECISIONS
USING GIS
AND REMOTE
SENSING

We hope this book helps build a bridge between your classroom and the centers of government, business, nongovernmental organizations, and other entities that rely on remote sensing and geographic information system (GIS) technology to make important decisions.

Making Spatial Decisions Using GIS and Remote Sensing: A Workbook puts actual spatial data into your hands—or more precisely, your computer—to analyze, interpret, and apply to various scenarios. You will make the types of decisions that affect an agency, a community, or a nation. The projects in this book focus on remote sensing but in the context of using GIS to do the analysis and make maps. They also help you hone your critical-thinking skills and pursue independent study. We have chosen scenarios we think are relevant, thought provoking, and applicable to a broad range of studies, not just geography. We also think you will enjoy finding solutions to challenging problems.

Making Spatial Decisions Using GIS and Remote Sensing comes with a Maps and Data DVD that contains the GIS data, maps, worksheets, and other documents you will need to complete the projects. The book also allows you to download ArcGIS 10.1 for Desktop software. This trial software requires the Microsoft Windows 7 operating system. The book also provides information about online resources.

Making Spatial Decisions Using GIS and Remote Sensing is a college-level textbook that presumes you have some prior GIS knowledge. It is an opportunity to learn about ways to connect remote sensing techniques and GIS analysis.

The technical prerequisites are outlined in the introduction, which also explains how to use the book, organize your workflow, and evaluate your work products. The introduction includes a summary of all 10 modules.

We have provided worksheets, available on the DVD, for your convenience. They will help you follow the activities by providing a place where you can log answers to questions along the way and keep track of the work to be completed. The answers to the individual questions will help you prepare your analysis to answer the larger problems in each module. Don't lose sight of the forest (the overarching problem) by spending too much time looking at the trees (the individual questions). Your instructor has supplemental resources available to assist you with the projects in this book.

We started our work in this area using raster-based imagery, and that work grabbed us and never let go. This book stems from our many decades of experience working with students and teachers at all levels. We hope it is an enjoyable learning experience for you!

Kathryn Keranen and Robert Kolvoord

ACKNOWLEDGMENTS

MAKING
SPATIAL
DECISIONS
USING GIS
AND REMOTE
SENSING

We would like to thank Dr. Zachary Bortolot of James Madison University and three reviewers of this book for their comments and feedback.

We would also like to thank the staff at Esri Press for all their assistance in bringing this book from concept to print.

Most important, we want to acknowledge the wonderful students and teachers in the Geospatial Semester with whom we get to work and who contributed to this project.

INTRODUCTION

MAKING
SPATIAL
DECISIONS
USING GIS
AND REMOTE
SENSING

GIS technology is a powerful way to analyze spatial data and is used in many industries to support decision making. From protecting fragile wetlands to mapping the impact of urban development, GIS is at the center of important decisions. In *Making Spatial Decisions Using GIS and Remote Sensing*, you will analyze remote sensing data and learn to make and support spatial decisions. This scenario-based book presents 10 modules, each with real-world situations designed to expand your remote sensing and GIS skills as you make decisions using actual data. The projects involve the complicated, sometimes messy, spatially related issues professionals from many disciplines face each day. The scenarios are more than step-by-step recipes for you to learn new remote sensing and GIS skills; they are opportunities to develop your critical-thinking prowess. As is true in so many situations, there is more than one "right answer," and you will have to decide the best solution for the problem at hand.

How to use this book

The scenario-based problems in this book presume that you have prior experience using GIS and are able to perform basic tasks using ArcGIS software (we are more specific about our expectations in the "Prior GIS experience" section of this introduction). Each of the 10 modules in this book follows the same format. Project 1 gives step-by-step instructions to explore a scenario. You will answer questions and complete charts and tables on a worksheet provided for you. You will then reach decisions to resolve the central problem and develop a presentation to share your solution. Project 2 provides a slightly different scenario and the requisite data without step-by-step directions. You will apply what you learned in project 1 and reinforce those skills and knowledge. Finally, each module includes an "On your own" project that suggests different scenarios you can explore by locating and downloading data and reproducing the analysis from the guided projects.

The modules in this book were designed to be done in order and ideally in conjunction with a remote sensing class. If you are working with these modules independently, you may find access to a remote sensing textbook very helpful. There are many good texts, including the following:

- Chuvieco, E., and A. Huete. 2010. *Fundamentals of Satellite Remote Sensing*. Boca Raton, FL: CRC Press.
- Jensen, J. R. 2004. *Introductory Digital Image Processing: A Remote Sensing Perspective*. Upper Saddle River, NJ: Prentice Hall.
- Lillesand, T., R. W. Kiefer, and J. Chipman. 2007. *Remote Sensing and Image Interpretation*. New York: Wiley.

- Mather, P. M., and M. Koch. 2011. *Computer Processing of Remotely-Sensed Image: An Introduction*. New York: Wiley.
- Weng, Q. 2012. *An Introduction to Contemporary Remote Sensing*. New York: McGraw-Hill.

MAKING
SPATIAL
DECISIONS
USING GIS
AND REMOTE
SENSING

GIS workflow

One major difference between these modules and other GIS-based lessons you may have done is the focus on GIS workflow: documenting and being systematic about the problem-solving process. Following a consistent GIS workflow is an important part of becoming a GIS professional.

We recommend the following GIS workflow:

1. Define the problem or scenario.
2. Identify the deliverables (mostly maps) needed to support the decision.
3. Identify, collect, organize, and examine the data needed to address the problem.
4. Document your work as follows:
   a. Create a process summary.
   b. Document your map.
   c. Set the environments.
5. Prepare your data.
6. Create a basemap or locational map.
7. Perform the remote sensing analysis.
8. Produce the deliverables, draw conclusions, and present the results.

A more detailed explanation is in the "Workflow" section of the book.

Process summary

The process summary is particularly important because it serves as a record of the steps you took in the analysis. It also allows others to repeat the analysis and verify or validate your results. A process summary might look something like this for an analysis of developed land in Virginia. It will, of course, vary for each project:

1. Examine the metadata.
2. Create a basemap.

Map document 1

1. Prepare a basemap of Virginia.
2. Symbolize counties in graduated color using the POP2000 field.
3. Label lakes and major rivers.
4. Prepare a presentation layout.
5. Save the map document.

Map document 2

1. Name the data frame percentage developed land.
2. Add the land-cover raster.
3. Symbolize the land-cover raster with unique values by land-cover type.
4. Measure/calculate the area of all land in Virginia.
5. Select developed land cover using the raster calculator.
6. Measure/calculate the area of developed land.
7. Calculate the percentage of developed land.
8. Prepare a presentation layout.
9. Save the map document.

The modules

The early modules in the book explore important aspects of the case study areas and reinforce key concepts in remote sensing, such as enhancement, creating color composite images, and spectral signatures. The latter modules focus on more specific issues, such as urban change, drought, and the Normalized Difference Vegetation Index (NDVI).

Chesapeake Bay (project 1)—modules 1–8

NASA Landsat satellites have played a significant role in the monitoring and management of the Chesapeake Bay watershed. Ongoing pollution problems have raised grave concerns about the continuing viability of the bay as a habitat and have severely curtailed the formerly lucrative fishing grounds (cf. Boesch 2006; Cresti, Srivastava, and Jung 2003; Kemp et al. 2005). The pollution comes from both urbanized areas and the rich agricultural lands upstream. Runoff from farms adds excess nitrates and phosphates to the bay. This often results in hypoxia (lack of oxygen in the water), leading to fish kill and depleting the bay's ability to sustain marine life. The size of the Chesapeake Bay (64,000 square miles) makes it difficult to get a comprehensive view of the bay and makes management decisions difficult. The Chesapeake Bay is one of the most highly monitored water bodies in the world, and Landsat imagery analysis provides a view of the Chesapeake at an appropriate scale to study land-cover changes and patterns.

MAKING
SPATIAL
DECISIONS
USING GIS
AND REMOTE
SENSING

INTRODUCTION

MAKING

SPATIAL

DECISIONS

USING GIS

AND REMOTE

SENSING

INTRODUCTION

The Landsat 5 image on the right was downloaded from the USGS Global Visualization Viewer (GloVis). It is an image centered on the portion of the Chesapeake Bay in Maryland and Virginia. The Landsat image is path 15/row 33, has 10 percent cloud cover, and is considered to be "leaf on," meaning that the trees have their foliage, because it was taken in May 2006.

In this project, your firm has been contracted by the Chesapeake Bay Foundation (CBF) to gather information about the Chesapeake Bay using Landsat imagery. Landsat imagery offers a view of the Chesapeake Bay at a scale that is appropriate for studying human population and land-use trends.

Decisions about the bay can be made using Landsat imagery and land-cover data derived from this imagery. The CBF has given your firm a very large contract to do this analysis. Your staff has decided to break the contract into eight different segments. Each segment will provide the CBF with a portion of analysis of the bay.

Las Vegas (project 2)—modules 1–8

Because of its desert climate, Las Vegas is particularly concerned about its rapid urbanization over the past 50 years. The natural vegetation around Las Vegas consists mainly of shrub/scrub, which is defined as areas dominated by shrubs less than 5 meters tall. It includes true shrubs and young trees that have been stunted by environmental conditions. This type of desert vegetation is fragile. The urban-growth trend has transformed this already fragile land cover by adding significant amounts of impervious land to the landscape (cf. Xian, Crane, and McMahon 2008). When studying Landsat imagery of the Las Vegas area, note that stone and crushed rock are used extensively as ground cover. Stone and rock have a spectral response similar to concrete. To the east of Las Vegas are Hoover Dam and the Lake Mead reservoir it has created. The Muddy and Virgin Rivers make up the top arm of Lake Mead, and you can also see Hoover Dam where the Colorado River leaves the reservoir. Landsat imagery has been used to map the receding Lake Mead shoreline caused by drought.

Your firm, Protect the Desert (PtD), has been hired by urban planners in Clark County, Nevada. The planners know that Landsat imagery offers a view of the Las Vegas area that is appropriate for use to gather information about the urban landscape and its encroachment on the desert environment. They want to use the Landsat imagery to map development intensity, look for patterns of urbanization, and discover other ways to use Landsat imagery as a decision-making tool. The Clark County urban planners have received a large federal grant to do this analysis and have awarded a substantial contract to PtD. Because of the tremendous scope of work, the PtD staff has decided to break the contract into eight different segments. Each segment will provide the planners with a portion of the analysis about Las Vegas and the surrounding area.

MAKING
SPATIAL
DECISIONS
USING GIS
AND REMOTE
SENSING

INTRODUCTION

Las Vegas Study Area

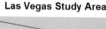

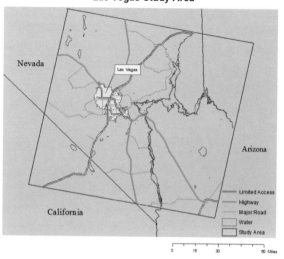

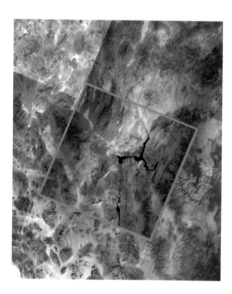

In project 2, a Landsat 5 image has been downloaded from GloVis. It is an image of Las Vegas, Nevada, that focuses on the city of Las Vegas. The Landsat image is path 39/row 35, has 10 percent cloud cover, and is considered to be leaf on because it was taken in May 2006.

Texas drought (project 1)—modules 9–10

Landsat imagery gives users a new way to study the omnipresent hazard of drought. Landsat imagery has been used at global, regional, and local scales to monitor drought. Shrinkage of water bodies and the effect of drought on different types of vegetation in the state of Texas is the focus of project 1 in modules 9 and 10. Texas has been experiencing drought conditions during recent years, and climatologists forecast more drought in the future. In module 9, Landsat imagery will be used to monitor the shrinkage of several lakes, and in module 10 you will use the NDVI to measure the effect of drought on different types of vegetation.

MAKING
SPATIAL
DECISIONS
USING GIS
AND REMOTE
SENSING

INTRODUCTION

These two Landsat 5 images were downloaded from GloVis. They are images of a section of Texas centered on San Angelo. The Landsat images are path 29/row 33. The left image was taken August 2, 2009, and the right image was taken August 11, 2011. Both images have 10 percent or less cloud cover.

Minnesota (project 2)—modules 9–10

Module 9, project 2, investigates Landsat band 6. Band 6 senses thermal infrared radiation. Its bandwidth is 10.4–12.5 micrometers (μ), and its resolution is 120 meters. It is the only Landsat band that can acquire night scenes. Band 6 can be used to detect temperatures of objects on Earth. Accurate temperatures of earth features can be determined only if the emissivity of an object is known. However, water has such high emissivity that the thermal band can accurately map surface water temperature. Water also has a very high capacity to store heat. Thermal band 6 has been used to monitor the water temperature of various lakes. In this module, Minnesota lake temperatures will be calculated and compared between June and November scenes.

Module 10, project 2, compares the NDVI of a June and a November scene centered on Minneapolis, Minnesota. NDVI of the complete scene will be compared, and NDVI of different types of vegetation will be individually investigated.

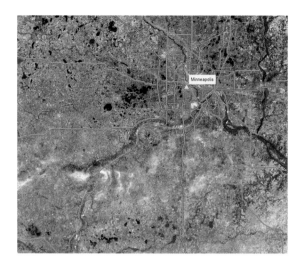

MAKING
SPATIAL
DECISIONS
USING GIS
AND REMOTE
SENSING

INTRODUCTION

These two Landsat 5 images were downloaded from GloVis. They are images of a section of Minnesota centered on Minneapolis. The Landsat images are path 27/row 29. The left image was taken June 1, 2009, and the right was taken November 11, 2010. Both images have 10 percent or less cloud cover.

Assessing your work

Your instructor will talk with you about assessment, but you can assess your own work before handing it in. The following items will help ensure your presentation maps are the best they can be.

Think about each of these items as you finish your maps and write up your work.

Map composition

Do your maps have the following elements?
- Title (addresses the major theme in your analysis)
- Legend
- Scale
- Compass rose
- Author (your name)

MAKING
SPATIAL
DECISIONS
USING GIS
AND REMOTE
SENSING

INTRODUCTION

Classification

Did you make reasonable choices for the classifications of the different layers in your maps? Is the symbology appropriate for the various layers?

- For quantitative data, is there a logical progression from low to high values, and are they clearly labeled?
- For qualitative data, did you make sure not to imply any ranking in your legend?

Scale and projection

- Is the map scale appropriate for the problem?
- Have you used an appropriate map projection?

Implied analysis

- Did you correctly interpret the color, pattern, and shape of your symbologies?
- Does any text you have written inform the reader of the map's intended use?

Design and aesthetics

- Are your maps visually balanced and attractive?
- Can you distinguish the various symbols for different layers in your maps?
- Are your maps accessible to all viewers (e.g., color-blind users)?

Effectiveness of map

- How well do the map components communicate the story of your map?
- Do the map components take into account the interests and expertise of the intended audience?
- Are the map components of appropriate size?

By thinking about these items as you produce maps and do your analysis, you will make effective maps that can be used to solve the problems in each module.

ArcGIS Resources at http://resources.arcgis.com provides support for making maps with ArcGIS software. The site assumes that the user has experience with ArcGIS software but may be a novice map maker with little or no experience in the field of cartography. The site is a great resource for tips on cartography and using ArcGIS to make high-quality maps.

Prior GIS experience

In these modules, we presume that you have used ArcGIS software before and that you can do the following basic tasks:

- Navigate and find data on local drives, on network drives, and on CDs and DVDs.
- Name files and save them to a known location.
- Use ArcCatalog to connect to a folder.
- Use ArcCatalog to preview a data layer and look at its metadata.
- Add data to ArcMap by dragging layers from ArcCatalog or using the Add Data button.
- Rearrange layers in the ArcMap Table of Contents.
- Identify the Table of Contents and the Map windows in ArcMap and know the purpose of each.
- Use the following tools:
  - Identify
  - Zoom in
  - Zoom out
  - Full extent
  - Pan
  - Find
  - Measure
- Symbolize a layer by category or quantity.
- Open the attribute table of a data layer.
- Select features by attribute.
- Label features.
- Select features on a map and clear a selection.
- Work with tables.
- Make a basic layout with map elements.

If you need some review, many great resources are available, including the GIS Tutorial series and *Getting to Know ArcGIS for Desktop* from Esri Press and ArcGIS 10.1 for Desktop Help online.

Setting up the software and data

The software and data to complete the exercises are provided with this book.

ArcGIS 10.1 for Desktop software

A 180-day trial version of ArcGIS 10.1 for Desktop Advanced license (single use) software can be downloaded at http://www.esri.com/MSDRemSenforArcGIS10-1. Use the code printed on the inside back cover of this book to access the download site, and follow the on-screen instructions to download and register the software.

MAKING
SPATIAL
DECISIONS
USING GIS
AND REMOTE
SENSING

INTRODUCTION

Please note that the 180-day trial of ArcGIS 10.1 for Desktop is limited to one-time use only for each software workbook. In addition, the ArcGIS for Desktop software trial is applicable only to new, unused software workbooks. The software trial cannot be reused or reinstalled, nor can the time limit on the software trial be extended.

MAKING
SPATIAL
DECISIONS
USING GIS
AND REMOTE
SENSING

The ArcGIS for Desktop software installation includes three ArcGIS for Desktop extension products used in this book: ArcGIS 3D Analyst, ArcGIS Network Analyst, and ArcGIS Spatial Analyst. ArcGIS 3D Analyst includes the ArcScene and ArcGlobe applications, which are used for three-dimensional visualization and exploration of geographic data. ArcGIS Network Analyst and ArcGIS Spatial Analyst provide tools for specialized analysis tasks.

All the help and resources you need to get up and running with your software are provided through videos, instructions, and commonly asked questions at http://resources.arcgis.com/en/help/.

Using the Maps and Data DVD

Refer to the installation guides at the back of the book for detailed system requirements and instructions on how to install the software data. The data license agreement is found at the back of the book and on the Maps and Data DVD. If you do not feel comfortable installing programs on your computer, please be sure to ask your campus technology specialist for assistance.

The software and data must be installed on the hard drive of all computers you will use to complete these modules. Installations on a computer network server are not recommended nor supported.

Metadata

Metadata (information about the data) is included for all the GIS data provided on the DVD. The metadata includes a description of the data, where it came from, a definition of each attribute field, and other useful information. This metadata can be viewed in ArcCatalog.

Troubleshooting ArcGIS

Instructions are written assuming the user interface and user preferences have the default settings. Unless you are working with a fresh installation of the software, however, chances are you will encounter some differences between the instructions and what you see on your screen. This is because ArcMap stores settings from a previous session. This could vary which toolbars are visible, where toolbars are located, the width of the table of contents, or whether or not the map scale changes when the window is resized.

WORKFLOW

Addressing and analyzing a remote sensing problem using GIS requires a structured approach similar to the problem-solving techniques you have used in other disciplines with other tools. By using this approach, you will be certain to develop a solution that can be shared with and repeated by other GIS users, and you will be able to communicate your results successfully to a broad audience.

MAKING
SPATIAL
DECISIONS
USING GIS
AND REMOTE
SENSING

As mentioned previously in this book, we use the following steps to define the remote sensing workflow or procedure. In each project, you will work through each step in order as you explore the problems and develop solutions or arrive at decisions. This is not the only way to work through a remote sensing problem, but it is widely used in practice. Here are the steps explored in detail in the following pages:

1. Define the problem or scenario.
2. Identify the deliverables needed to support the decision.
3. Identify, collect, organize, and examine the data needed to address the problem.
4. Document your work.
5. Prepare your data.
6. Create a basemap or locational map.
7. Perform the remote sensing analysis.
8. Produce the deliverables, draw conclusions, and present the results.

1. Define the problem or scenario

This is perhaps the most difficult part of the entire process. You must define what issue you are trying to address from a complicated sea of information and perhaps competing interests.

This book presents real-world scenarios to help you with this process. You should always try to focus on the core issue in any situation that calls for remote sensing analysis. What decision needs to be made? Who is going to make it? What do people need to know to make a rational decision?

One approach is to write down a short description of the problem, including the general scenario, the stakeholders, and the specific issues that need to be addressed. It also helps to think about what decisions will ultimately be made using the data. Of course, the scenarios in this book require remote sensing analysis using ArcGIS software to solve problems. Other kinds of analyses may be included in these scenarios, but the focus will be on remote sensing problem solving.

2. Identify the deliverables needed to support the decision

After you have defined the problem, you need to think about what maps and other visualizations and descriptions you will produce to help you analyze and solve the problem. These may include the following:

- Maps
- Charts
- Tables of calculations
- Written analysis

MAKING
SPATIAL
DECISIONS
USING GIS
AND REMOTE
SENSING

WORKFLOW

By identifying these visualizations and descriptions, you will be able to identify the data required for your analysis. You will also be able to determine whether you have defined your problem in sufficient detail to develop a solution. Along with specifying your data requirements, your list of deliverables will also guide your analysis. These first steps of your workflow are very iterative. You may need to go through them a few times before you feel ready to begin your analysis. In many projects, they are also the most complicated steps.

3. Identify, collect, organize, and examine the data needed to address the problem

Once you have defined your problem and identified the deliverables, it is time to search for data. As a starting point, you should identify data for a basemap and imagery to solve the problem you have defined. In many instances, you will have data at hand that will allow you to pursue your analysis. This data may have been provided for you (as in the first two projects of each module), or it may be part of a collection of data where you are working. The data may come from the Data and Maps for ArcGIS or from ArcGIS Online. However, in some instances you may need to get data from other sources, such as the Internet, or you may need to collect your own data. If you need to do some sort of field study to collect data, you will want to be sure to develop a clear protocol for acquiring data and follow it consistently.

You should be careful to identify your data sources and make sure you have permission to use the data for your particular problem. Make sure all data layers and imagery have appropriate metadata that describes various aspects of the data, including the creator of the data, the map projection, and the attributes included. You will also want to know the accuracy or resolution of each layer.

As you collect the data for your analysis, we strongly recommend you adopt a standard for how you organize and store this data. This organization will make your analysis much easier, and you will be able to quickly find different layers and share your work with others.

A standard directory structure looks like this:
- Project folder
    - Data folder containing all data layers
    - Document folder containing all project documentation
    - Project.mxd (the ArcGIS map document)

MAKING
SPATIAL
DECISIONS
USING GIS
AND REMOTE
SENSING

WORKFLOW

You must be able to both read and write to all folders within the project folder. You can place the project folder at a convenient location in the network structure in which you work. Each module shows you how to save your data with relative paths to make data sharing even easier.

When you obtain data from other sources, you will often want to take a quick peek at the data to make sure you understand what the data actually represents. As we mentioned earlier, these first steps of your workflow are very iterative. You may need to go through them a few times before you feel ready to begin your analysis. In some instances, you may need to adjust the definition of the problem to let you use the available data, or the deliverables may need to be modified to make the analysis possible.

4. Document your work

Documenting your project and creating a process summary is critical to keeping track of the various steps in your analysis. To document your project, go to the File menu and click Document Properties. In that dialog box, you can enter some of the basic information about your project. A process summary is simply a text document that keeps track of the different steps you use in your analysis. Too often, documentation of GIS work is left to the end of the project, or it is not done at all. We encourage you to start your process summary early in each project and to keep up with it as you proceed.

5. Prepare your data

Accurate remote sensing analysis with GIS may call for changing the units in which measurements will be made. You will also want to know the units in which various quantities are measured; for example, are the elevations in feet or meters above sea level? You may need to provide geographic references to certain quantities, such as adding GPS-based data to your map display. You will be guided through these steps in the different projects.

6. Create a basemap or locational map

Finally, you are ready to make maps and perform your remote sensing analysis. Your first step in this process should always be to build a locational map or basemap that shows the area you are studying. A basemap will typically contain the major features of the area, such as roads and streams, and it will help orient you geographically to the area and its features. This is a good practice when solving any remote sensing problem, because it will give you a sense of the scale of your study area and the different features that may dominate the area.

7. Perform the remote sensing analysis

Now it is time to get down to solving the problem. In this step, you will apply the different remote sensing tools to the data you have compiled. The point of your analysis is to produce the deliverables that you specified initially and allow you to develop a solution to the problem you defined. Often you will find that the first set of analysis tools may not provide results that help solve the problem, and you will need to refine your analysis and try other tools or techniques. Even if you have been very careful and thorough in your planning to this point, geographic data and imagery never lose their ability to surprise you. As you work through the modules in this book, you will learn a variety of advanced analysis techniques.

8. Produce the deliverables, draw conclusions, and present the results

Finally, you are satisfied with your analysis and ready to complete your work. You will first need to complete your deliverables (as defined earlier in the process). This may mean making map layouts, graphs, charts, or tables. You will also want to finish documenting your analysis process in the process summary. Remember to keep the principles of good cartographic design in mind when you make your deliverables. Remember, ArcGIS Resources at http://resources.arcgis.com is an excellent resource for cartographic information.

You will also need to write a report that states your conclusions and justifies them, using the deliverables you have produced. Always keep the audience in mind as you prepare to report your results. A technically savvy audience will have very different needs from a group of high-level decision makers.

MAKING
SPATIAL
DECISIONS
USING GIS
AND REMOTE
SENSING

WORKFLOW

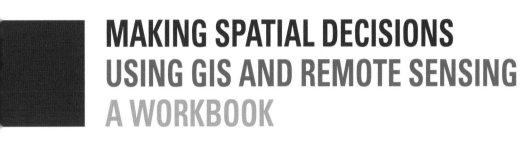

MAKING SPATIAL DECISIONS
USING GIS AND REMOTE SENSING
A WORKBOOK

MODULE 1
ENHANCING IMAGES

Introduction

To effectively use imagery to analyze and solve problems, you need to be able to display and enhance imagery and interpret pixel values. In this module, you will create basemaps of the study area and explore the basic functionality of the Image Analysis window in ArcMap. You will enhance and display imagery in a variety of ways and interpret the results.

Scenarios in this module

- Displaying and enhancing Landsat imagery of the Chesapeake Bay
- Displaying and enhancing Landsat imagery of Las Vegas, Nevada
- On your own

Student worksheets

The student worksheet files can be found on the Maps and Data DVD.

Project 1: Chesapeake Bay student sheet
- File name: 01a_enhance_worksheet
- Location: \Student_Worksheets\01enhance

Project 2: Las Vegas student sheet
- File name: 01b_enhance_worksheet
- Location: \Student_Worksheets\01enhance

MAKING
SPATIAL
DECISIONS
USING GIS
AND REMOTE
SENSING

1

*ENHANCING
IMAGES*

PROJECT 1
Displaying and enhancing Landsat imagery of the Chesapeake Bay

Background

Landsat Thematic Mapper (TM) images consist of seven spectral bands with a spatial resolution of 30 meters for bands 1–5 and band 7. Band 6 is a thermal infrared band that has a spatial resolution of 120 meters. Table 1 provides information about these bands.

TABLE 1 DESCRIPTION OF LANDSAT SPECTRAL BANDS

Band	Wavelength	Micrometers	Resolution (meters)	Band focus (features most clearly identifiable in a band)
1	blue	0.45–0.52	30	Coastal water (has the most atmospheric scatter)
2	green	0.52–0.60	30	Vegetation
3	red	0.63–0.69	30	Chlorophyll absorption (monitors vegetation health)
4	near-infrared	0.76–0.90	30	Water discrimination (water appears very dark)
5	mid-infrared	1.55–1.75	30	Moisture content of vegetation and soil
6	thermal	10.40–12.50	120	Surface temperature, volcanic monitoring
7	far-infrared	2.08–2.35	30	Rock discrimination

The individual band images are stored using digital numbers (DN values) between 0 and 255. For a single band, it is common to display the different numbers using shades of gray.

DATA (Current Workspace) \01enhance\project1_bay_enhance\bay_data_enhance
RESULTS (Scratch Workspace) \01enhance\project1_bay_enhance\bay_results_enhance

Most computer monitors are capable of displaying 256 distinct shades of gray. In bands 1–5 and 7, the lower the value of the pixel in the grayscale image, the darker the pixel, and the lower the spectral reflectivity in that area. In band 6, lower pixel values and darker shades of gray approximately correspond to lower temperatures. Water, which absorbs near-infrared radiation, has a low reflectivity/digital number in band 4. Concrete, sand, and other man-made surfaces have high reflectivity in certain bands, and therefore have high digital numbers.

The purpose of image enhancement is to use different procedures to improve the visual appearance of an image, making it easier to interpret. Image enhancement can occur via a number of methods. Each method tries to use the full range of shades of gray that can be displayed by the computer monitor. This is important because in many images the pixels have similar DN values and do not cover the full range of possible values. With the image displayed using the full range of brightness, differences can be more readily observed because they are amplified in the enhanced display.

MAKING
SPATIAL
DECISIONS
USING GIS
AND REMOTE
SENSING

ENHANCING
IMAGES

Scenario/problem

Because visually interpreting Landsat images is important in many qualitative applications, your company has designated the first segment of the contract to understanding and enhancing grayscale Landsat images. The Chesapeake Bay Foundation (CBF) has asked for enhanced grayscale imagery it can use in articles and presentations to highlight different aspects of the bay.

Objectives

You have been asked to use Landsat imagery to do the following:
- Display the different bands of Landsat 5 data.
- Investigate histograms and their relationship to spectral reflectance.
- Use different enhancement techniques for grayscale imagery to highlight important features in the Chesapeake Bay.

Deliverables

We recommend the following deliverables for this project:
1. A map that displays and identifies the Chesapeake watershed, the study area, the rivers, and the highways.
2. A discussion of what bands to use to best show water, roads and highways, and different types of vegetation.
3. Histograms comparing reflectance values for bands 1–5 and 7.

DATA (Current Workspace)   \01enhance\project1_bay_enhance\bay_data_enhance
RESULTS (Scratch Workspace)   \01enhance\project1_bay_enhance\bay_results_enhance

5

4. A document with three different enhanced images with their respective histograms and a written explanation of the enhancement techniques used.

The questions for this project are both quantitative and qualitative. They identify key points that should be addressed in your analysis and presentation.

Keeping track of where your data and results are located is always a challenge. In these projects, we give the path to access the data (current workspace) and the path to store the results (scratch workspace) in a footnote on each page. The directions will specify whether results should be placed in the results folder or the results geodatabase.

MAKING
SPATIAL
DECISIONS
USING GIS
AND REMOTE
SENSING

ENHANCING
IMAGES

Examine the data

The next step in your workflow is to identify, collect, and examine the data for the Chesapeake Bay analysis. Here, we have identified and collected the data layers you will need. Explore the data to better understand both the raster and vector feature classes in this project. You can do this by right-clicking each feature class or raster within the geodatabase and clicking Item Description. This can be done in ArcMap using the Catalog window without actually adding the layers to the map document. The item description is a simple, standardized core description that contains a title, summary, tags, credits, and restrictions. If you want to more thoroughly access the metadata associated with each layer, you will need to perform the following steps:

1. On the Customize menu, click ArcMap Options.
2. Click the Metadata tab, and then select ISO 19139 Metadata Implementation Specification from the drop-down menu.

You have to do this only once, and then it is set as the standard. Now formal metadata is supported in your item description. This allows you to view the spatial coordinate system, the resolution of the data, and the attribute data in the ArcGIS metadata.

You are now ready to examine the metadata.

1. Start ArcMap. (For these exercises, the Getting Started dialog box is not needed. Select the check box at the bottom so it does not appear in the future.)

2. Connect to the geodatabase in the data folder. Remember that the path to the data folder is shown in the footnote.

3. Right-click rivers in the Catalog window and view Item Description.

4. Open the landsat_may_2006 folder and double-click the L5015033_03320060504_MTL text file to access the metadata for the Landsat image.

Q1 *View the metadata for these features and complete the following chart on your worksheet.*

Layer	Publication information: Who created the data?	Time period data is relevant	Spatial horizontal coordinate system	Data type	Resolution for Rasters
L5015033_03320060504_B40					
Rivers					

MAKING
SPATIAL
DECISIONS
USING GIS
AND REMOTE
SENSING

ENHANCING
IMAGES

Now that you have explored the available data, you are almost ready to begin your analysis. First you need to start a process summary, document your project, and set the project environments.

Organize and document your work

The following preliminary steps are essential to a successful geospatial analysis.

Examine the directory structure

In a geospatial project, you must carefully keep track of the data and your calculations. You will work with a number of different files, and it is important to keep them organized so you can easily find them. The best way to do this is to have a folder for your project that contains a data folder. For this project, the folder named **\01enhance\project1_bay_enhance\bay_data_enhance** will be your project data folder. Make sure it is stored in a place where you have write access.

You can store your data inside the results folder. The results folder already contains an empty geodatabase named **bay_results** for this purpose. Save your map documents to the **\01enhance\ project1_bay_enhance\bay_results_enhance** folder.

Folder structure:
01enhance
        project1_bay_enhance
            bay_data_enhance
                bay.gdb
                landsat_may_2006
            bay_results_enhance
                bay_results.gdb

Create a process summary

The process summary is simply a list of the steps you used to perform your analysis. We suggest using a simple text document for your process summary. Keep adding to it as you do your work to avoid forgetting any steps. The following list shows an example of the first few entries in a process summary:

1. Explore the data.
2. Produce a map of the Chesapeake Bay with the states and rivers labeled.
3. Display individual Landsat bands with histograms.

A process summary is essentially a work log. It will allow you to remember the steps in your data analysis at a later time or communicate them with others who might need to reproduce your work. When performing any sort of remote sensing or GIS analysis, you will often need to experiment to find the best methodology for completing a task. Be sure to revise your process summary when you are done with your analysis.

Document the map

You need to add descriptive properties to every map document you produce. You can use the same descriptive properties for each map document in the project or individualize the documentation from map to map.

1. On the File menu, click Map Document Properties. The Document Properties dialog box allows you to add a title, summary, description, author, credits, tags, and hyperlink base.

2. Click the pathnames check box, which makes ArcMap store relative pathnames to all your data sources. Storing relative pathnames allows ArcMap to automatically find all relevant data if you move the project folder to a new location or computer.

Set the environments

In GIS analysis, you will often gather data from several sources, and this data may be in different coordinate systems and map projections. When using GIS to perform area calculations, you would like your result to be in familiar units, such as square miles or square kilometers. Data in an unprojected geographic coordinate system has units of decimal degrees, which is difficult to interpret. Thus, calculations will be more accurate if all the feature classes involved are in the same map projection. Fortunately, ArcMap can do much of this work for you if you set certain environment variables and data frame properties. In this section, you will learn how to make these settings. To display your data correctly, you will need to set the coordinate system for the data frame. When you add data with a defined coordinate system, ArcMap will automatically set

MAKING
SPATIAL
DECISIONS
USING GIS
AND REMOTE
SENSING

ENHANCING
IMAGES

DATA (Current Workspace)   \01enhance\project1_bay_enhance\bay_data_enhance
RESULTS (Scratch Workspace)   \01enhance\project1_bay_enhance\bay_results_enhance

the data frame's projection to match the data. If you add additional layers that have coordinate systems that differ from the data frame's, they will automatically be reprojected on the fly to the data frame's coordinate system.

MAKING
SPATIAL
DECISIONS
USING GIS
AND REMOTE
SENSING

1. On the View menu, click Data Frame Properties, and then click the Coordinate System tab. Set the map projection to Projected Coordinate Systems > UTM > WGS 1984 > Northern Hemisphere > WGS 1984 UTM Zone 18N.

2. On the Geoprocessing menu, click Environments. Remember: The environment settings apply to all the functions within the project. The analysis environment includes the workspace where the results will be placed and the extent, cell size, and coordinate system for the results.

3. By default, inputs and outputs are placed in your current workspace, but you can redirect the output to another workspace, such as your results folder.
   a. Set the Current Workspace to \01enhance\project1_bay_enhance\bay_data_enhance.
   b. Set the Scratch Workspace to \01enhance\project1_bay_enhance\bay_results_enhance\ bay_results.

4. For Output Coordinate System, select Same as Display.

5. Click OK and save the project as enhance_bay1.

Analysis

Once you have examined the data, completed the map documentation, and set the environments, you are ready to begin the analysis and complete the data displays you need to explore the problem. A good place to start any geospatial analysis is to produce a basemap to better understand the distribution of features in the geographic area you are studying. First, you will prepare a basemap of the Chesapeake Bay, labeling the states and rivers.

STEP 1: Create a basemap of the Chesapeake Bay

1. Create a basemap of the Chesapeake Bay by adding highways, rivers, states, and watershed from the bayfeatures dataset in the bay geodatabase. Highways and rivers have been clipped to the area near Washington, DC.

2. Make the Chesapeake Bay watershed hollow and label the states. Convert the state labels to annotations within the map so that they can be moved and properly placed.

DATA (Current Workspace)   \01enhance\project1_bay_enhance\bay_data_enhance
RESULTS (Scratch Workspace)   \01enhance\project1_bay_enhance\bay_results_enhance

9

3. Symbolize the rivers by unique value using Name as the value field.

4. Add highways and import the highways layer file.

5. Make the study area 65% transparent.

MAKING
SPATIAL
DECISIONS
USING GIS
AND REMOTE
SENSING

6. Create a presentation-quality map using correct cartographic principles.

7. Save your map document to the bay_results_enhance folder.

8. Save the map document a second time as enhance_bay2. (When you save the map document again as enhance_bay2, it correctly saves the documentation, data frame projection, and environment settings. This saves you from having to redo these variables for subsequent deliverables.)

Q2 *What state has the most coastline on the Chesapeake Bay?*

Q3 *Write a short description of the geography of the Chesapeake Bay.*

Deliverable 1: A map that displays and identifies the Chesapeake watershed, the rivers, and the highways.

STEP 2: Investigate individual grayscale bands

1. Open the map document enhance_bay2 and go to Data View.

2. Remove all the layers except states. Symbolize the states as hollow.

3. Select all elements and remove the annotated labels.

4. Add L5015033_03320060504_B40 from the landsat_may_2006 folder in your data folder.

5. On the Windows menu, click Image Analysis to open the window.

6. Click the L5015033_03320060504_B40 image in the Image Analysis window and select the Background check box. This will remove the area that has no data.

7. Move the hollow states layer above the image.

8. Explore the image. Activate the Go To XY tool on the Tools toolbar. Change the units to meters using the drop-down menu and enter X: 323168 and Y: 4305831. Click the Zoom To icon and zoom to the Washington, DC, area. Locate the following features:

- Washington Mall
- Lincoln Memorial Reflecting Pool
- Jefferson Memorial

MAKING
SPATIAL
DECISIONS
USING GIS
AND REMOTE
SENSING

You can add different basemaps using the Add Data button on the Standard toolbar. There are a variety of basemaps to choose from (such as World imagery). Use the Effects toolbar to swipe or flicker to compare the Landsat image to the basemap imagery.

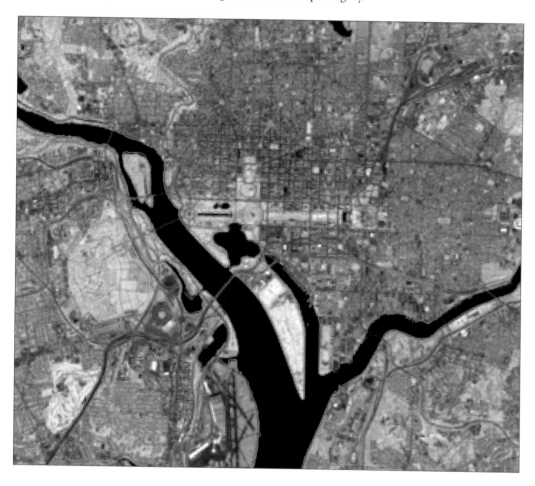

Q4 *What features can you identify?*

Q5 *What is the smallest feature you can identify? Why?*

MAKING
SPATIAL
DECISIONS
USING GIS
AND REMOTE
SENSING

1

ENHANCING
IMAGES

Landsat TM cannot detect small objects such as trees or individual houses. It can, however, detect larger types of land use, such as forest, agriculture, or urban areas. The "on the ground" resolution of each pixel is 30 × 30 meters (m), which means features smaller than 30 m, such as individual houses and cars, can't be resolved in these images. Although images with higher resolution are often available, they are very difficult to use for examining medium to large areas because of their very large file size. Landsat TM images are ideal for this purpose because of their lower resolution and smaller file size.

Images can be enhanced by using the tools available on the Image Analysis Display panel. This panel can be activated from the Windows menu and has the following tools:

- Contrast: Adjusts the difference between the darkest and lightest colors.
- Brightness: Increases the overall brightness of the image—for example, making dark colors lighter and light colors whiter.
- Transparency: Allows you to make a raster layer partially transparent.
- DRA (Dynamic Range Adjustment): Allows you to see features in some images or raster datasets, or to distinguish them more easily from their surroundings. In many cases, you can get the enhancement you need by adjusting the pixel values within the portion of the raster displayed and not using all the pixels in the full raster dataset.
- Background: Uses values to set the display background color.

You used the Background tool when you selected the Background check box and set the no-data values to no color.

Stay zoomed to Washington, DC, and experiment with the enhancement tools. To reset the image to its original value, just click the icon to the left of the tools.

9. Click the band and move the Contrast slider back and forth.

Q6 *What happens to the image when you change the contrast?*

10. Reset the Contrast tool and move the Brightness slider back and forth.

Q7 *What happens to the image when you change the brightness?*

DATA (Current Workspace)   \01enhance\project1_bay_enhance\bay_data_enhance
RESULTS (Scratch Workspace)   \01enhance\project1_bay_enhance\bay_results_enhance

11. Reset the Brightness tool and move the Transparency slider back and forth.

Q8   ***What happens to the image when you change the transparency?***

12. Zoom to the Washington Mall and select the DRA check box.

MAKING
SPATIAL
DECISIONS
USING GIS
AND REMOTE
SENSING

Q9   ***When you select the DRA check box, what happens to the features around the Lincoln and Jefferson Memorials?***

13. Clear the Background and DRA check boxes.

14. Add all the bands from the Landsat folder except L5015033_03320060504_B60. Arrange the bands in numerical order (B10 = band 1, B20 = band 2, and so on).

15. Add AOI (areas of interest) from the bayfeatures dataset.

ENHANCING
IMAGES

16. Make AOI hollow and label the features.

17. Zoom to AOI area 7. Start with the first Landsat band (B10) and turn the different bands on and off.

Q10   ***In which band are the bridges and streets more visible?***

Q11   ***Which band best delineates water?***

18. Zoom to AOI area 4.

Q12   ***Which band best distinguishes different types of vegetation?***

Deliverable 2: A discussion of which bands best show water, roads and highways, and different vegetation varieties.

DATA (Current Workspace)   \01enhance\project1_bay_enhance\bay_data_enhance
RESULTS (Scratch Workspace)   \01enhance\project1_bay_enhance\bay_results_enhance

13

STEP 3: Match DN to land features

1. Highlight all the bands in the Image Analysis Display panel, and then click Background to set the background values to no color.

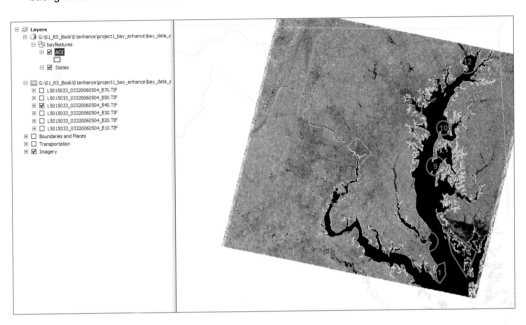

MAKING
SPATIAL
DECISIONS
USING GIS
AND REMOTE
SENSING

1

ENHANCING
IMAGES

Remember: We are not adding L5015033_03320060504_B60 because L5015033_03320060504_B60 shows emitted thermal radiation and not spectral reflectance.

In Landsat imagery, the attribute table contains a Value field, which represents the amount of reflected light (given as the DN), and a Count field, which gives the number of pixels that have that value. A raster attribute table must be calculated for each band before it is viewed. If a Landsat band has already been viewed, the attribute table has already been calculated.

If the attribute table has not been calculated, click the band and go to Properties > Symbology > Unique Values. The software will ask whether you want to build an attribute table. Click Yes and close the menu. Now you can open the attribute table for the band. Change the classification back to Stretched in the Show window to change the view back to gray. Click OK. An attribute table must be calculated for each of the six bands.

The digital numbers shown in the attribute table describe the brightness of features in different bands. Remember, in the bands you have viewed, the brighter (more reflective) a feature is, the higher its digital number. Consequently, if you create histograms that plot the frequency of the various digital numbers, different scenes will yield very different histograms. For instance, if you have a scene with a lot of water, the histogram for band 4 will have many pixels with low digital numbers (toward the left side of the histogram) because water is not very reflective in the near-infrared. However, a scene with significant forest cover would have a band 4 histogram with many pixels having high digital numbers because vegetation is very reflective in this band.

MAKING
SPATIAL
DECISIONS
USING GIS
AND REMOTE
SENSING

2. You can analyze the bands by identifying and selecting features by their digital number (reflectivity). For example, if you use the Identify tool you can see that water is usually in the range of 8–19 in band 4 (B40) rasters.

For the rest of the project, the Landsat bands will be referred to as bands 1–5 and 7.

3. If you select these values in the attribute table, pixels with these values will be highlighted in the image. Right-click Count in the attribute table to sort the table by ascending or descending values.

4. You can also use the Identify tool to click individual pixels that you know represent the land feature you are trying to identify. This will give you a starting value to investigate. The following image shows an example of selecting the DN values for water.

DATA (Current Workspace)   \01enhance\project1_bay_enhance\bay_data_enhance
RESULTS (Scratch Workspace)   \01enhance\project1_bay_enhance\bay_results_enhance

15

Q13 *Using the Identify tool, complete the following chart on your worksheet.*

| | Range of digital numbers | | | | | |
	Band 1	Band 2	Band 3	Band 4	Band 5	Band 7
water						

MAKING
SPATIAL
DECISIONS
USING GIS
AND REMOTE
SENSING

1

ENHANCING
IMAGES

STEP 4: Use the Pixel Inspector

The Pixel Inspector tool is used to view the pixel values of your raster dataset. It is different from the Identify tool. The Pixel Inspector will show an array of pixels instead of a single pixel value. It is useful in examining areas where pixel values change abruptly. The Pixel Inspector tool has to be added to a toolbar.

1. In ArcMap, go to Customize > Customize Mode > Commands > Categories > Raster.

2. From the Commands list, drag the Pixel Inspector onto any toolbar.

3. Click Close.

Once again, a good place to look at different pixel values is the Washington, DC, National Mall area.

4. Zoom to the National Mall area.

16

DATA (Current Workspace)   \01enhance\project1_bay_enhance\bay_data_enhance
RESULTS (Scratch Workspace)   \01enhance\project1_bay_enhance\bay_results_enhance

5. When you click a pixel with the Pixel Inspector, a window opens with the pixel value displayed in the upper-left corner. The pixel value changes as you pick other areas. The query area is also shown as a grid in another display window.

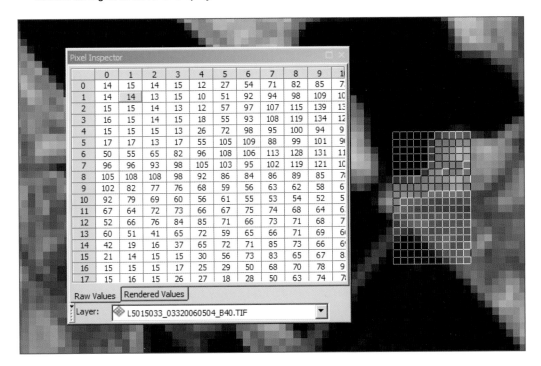

MAKING
SPATIAL
DECISIONS
USING GIS
AND REMOTE
SENSING

ENHANCING
IMAGES

Q14 *How do the values of the pixels change as you move from water to land?*

Q15 *Do the pixel values change as you go over the bridges going into DC?*

6. Save your map document to your results folder.

7. Save the map document again as enhance_bay3.

DATA (Current Workspace)   \01enhance\project1_bay_enhance\bay_data_enhance
RESULTS (Scratch Workspace)   \01enhance\project1_bay_enhance\bay_results_enhance

17

STEP 5: Create frequency histograms and an all band image frequency histogram

Frequency histograms provide another way to explore the reflectivity distribution of the bands, using a graph showing the variation in each band's reflectivity. A frequency histogram shows the values of the pixels (x-axis) in a raster plotted versus the number of pixels with that value (the frequency) on the y-axis. For example, as mentioned earlier, in band 4 low reflectivities, or values, are an indication of water, and high values indicate features with high reflectivity (e.g., concrete, sand).

MAKING
SPATIAL
DECISIONS
USING GIS
AND REMOTE
SENSING

1. Open the map document enhance_bay3.

2. In the Image Analysis window, clear the Background check box for all bands.

3. On the View menu, go to Graphs > Create Graph.
   a. Create a vertical bar graph of L5015033_03320060504_B40.
   b. The value field is Count.
   c. Expand the graph so that you can read the full x- and y-axis values.

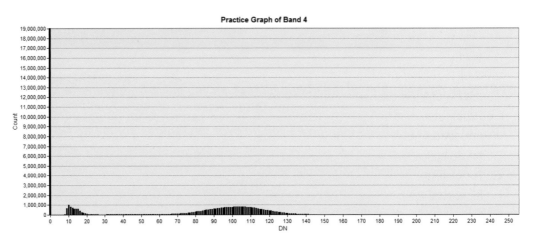

Q16 *What is the total number of pixels in the image?*

Hint: You might want to open the attribute table. Remember, you can right-click Count, and then click Statistics.

Q17 *How many pixels have a DN of 100?*

Q18 *How many pixels have a DN of approximately 10?*

Q19 *How would you describe the graph?*

DATA (Current Workspace)   \01enhance\project1_bay_enhance\bay_data_enhance
RESULTS (Scratch Workspace)   \01enhance\project1_bay_enhance\bay_results_enhance

Q20 *What does the large number of pixels with a DN of 0 represent?*

Now that you understand the basics of a frequency histogram, you need to make a composite histogram showing all six nonthermal Landsat TM bands.

4. Close the practice graph for band 4 and go to View Graph > Manage and delete the practice graph. You are now ready to graph the frequency histogram for each band.

You do not want to graph the background values. Open the attribute table and select the value 0, which is the background value. Switch the selection to select the other values. This has to be done for each band or every time you graph a new series.

Do this procedure for each band before you start making the histograms.

5. On the View menu, click Create Graph. Create a vertical bar graph of band 1 with the following properties:
    a. Set the Value field to Count.
    b. For color, click Custom and choose a color appropriate to the band. For band 1, select blue.
    c. The X label field should be Value.
    d. Click Next, and then click the option that says, "Show only selected features/records on the graph."
    e. Click Back, and then click Add a New Series. Graph band 2. Pick an appropriate color. Do not forget to exclude the background values by showing only the selected features.
    f. Repeat steps a–e for the other bands. Hint: At the bottom of the Graph window, you will see tabs for the graph for each band. If you click the tab once, you can change its name. Turn on the 3D option.
    g. Click Finish.
    h. Right-click the title to access Advanced Properties.
    i. Under Advanced Properties, click Axis/Left Axis and change the maximum to 4,000,000.

MAKING
SPATIAL
DECISIONS
USING GIS
AND REMOTE
SENSING

ENHANCING
IMAGES

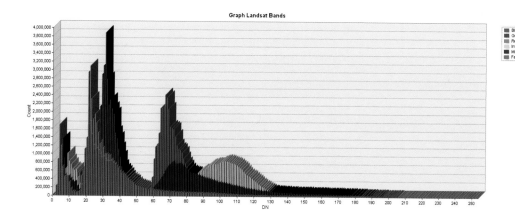

DATA (Current Workspace)   \01enhance\project1_bay_enhance\bay_data_enhance
RESULTS (Scratch Workspace)   \01enhance\project1_bay_enhance\bay_results_enhance

19

Q21 Which bands have histograms with bimodal distribution?

Q22 Which band has the most compact distribution of reflectance?

Q23 Which band has the broadest distribution of reflectance?

MAKING
SPATIAL
DECISIONS
USING GIS
AND REMOTE
SENSING

ENHANCING
IMAGES

6. Save your map document to your results folder.

7. Save the map document again as enhance_bay4.

Deliverable 3: Histogram chart comparing reflectance values for bands 1–5 and 7.

STEP 6: Enhance images

Image enhancement involves procedures that improve the appearance or interpretability of an image. As previously stated, a single image (band) is visualized in shades of gray, ranging from black (0) to white (255). To produce an image that has an optimal contrast, the full range of values from 0 to 255 should be used. There are various methods for contrast enhancement. In this exercise, you will look at three of these methods. In each of these methods, changes are not made to the actual data, but only to the way in which the values in the image are mapped to the values of the display.

The default method in ArcMap is the two standard deviations stretch, and it is used to brighten raster images. The minimum-maximum stretch is used when you need to display an image with a narrow range of pixel values. Unlike the previous two methods, the histogram equalization stretch lets you control the range of values displayed. These techniques are generally referred to as "stretching" an image.

1. Open the map document enhance_bay4.

2. Delete the graph.

3. Clear all selections.

4. Remove all layers except band4.

DATA (Current Workspace)   \01enhance\project1_bay_enhance\bay_data_enhance
RESULTS (Scratch Workspace)   \01enhance\project1_bay_enhance\bay_results_enhance

5. Right-click band4 and go to Properties > Symbology. The classification type is stretched, and you can see the default type is Standard Deviations with an n-value of 2.

6. Click the Histograms button, and the histogram will be displayed. The gray values are the original pixel values, and the purple values are the "displayed" values after the stretch. The line on the histogram shows the relationship between the input values and the output values. A straight line represents a linear stretch, and a curved line represents a nonlinear stretch.

MAKING
SPATIAL
DECISIONS
USING GIS
AND REMOTE
SENSING

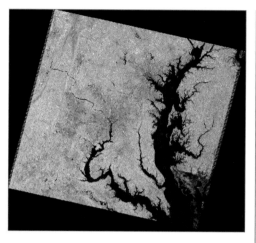

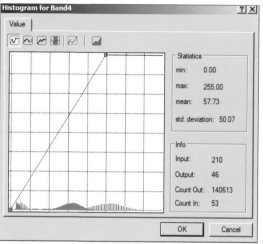

Q24 *Describe how the pixels are distributed after the standard deviation stretch.*

Q25 *Explain the initial spike on the left side of the graph.*

If you eliminate the background value of 0, the graph represents the distribution of spectral reflectance values in the image more accurately.

7. Remove band4 and add clip_band4 from the bay geodatabase.

Q26 *How does clip_band4 look different from band4?*

8. Look at the histogram again.

Q27 *Why does the histogram look different?*

Q28 *What is the minimum value of the image?*

MAKING
SPATIAL
DECISIONS
USING GIS
AND REMOTE
SENSING

1

ENHANCING

IMAGES

9. Capture a screen shot of the image and the histogram using either the Print Screen key on the keyboard or any available image capture software. Insert both images into your document.

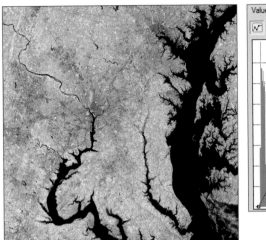 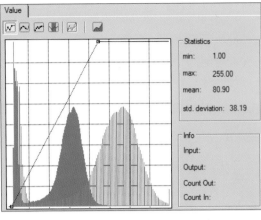

10. The preceding histogram used the standard deviation stretch. Repeat directions 8 and 9 and select the histogram equalization stretch with clip_band4.

Be sure to click Apply when you apply the histogram equalization stretch.

Q29 *How are the pixels distributed when a histogram equalization stretch is used?*

11. Capture a screen shot of the image and the histogram. Insert both into your document.

12. Repeat directions 8 and 9 for a minimum-maximum stretch with clip_band4. In this stretch, you have to set the minimum and maximum values manually. Click the small box on the x-axis and drag it either to the origin or the lowest pixel value you want to include. This sets the minimum. Now click the small box at the top and drag it to the highest pixel value you want to include. This sets the maximum. When you examine the histogram, you will see that the majority of the pixels are in the middle. Very few pixels have the minimum or maximum values. It is common practice in a minimum-maximum stretch to cut off the tails of the distribution by putting them outside the maximum and minimum values you have chosen.

13. Capture a screen shot of the image and the histogram. Insert both into your document.

Q30 *Compare the appearance of an image with a histogram equalization stretch to an image with standard deviation or minimum-maximum stretch.*

14. Save your map document to your results folder.

Deliverable 4: A document with the three different enhanced images with their respective histograms and a written comparison of the enhancement techniques.

MAKING
SPATIAL
DECISIONS
USING GIS
AND REMOTE
SENSING

ENHANCING
IMAGES

DATA (Current Workspace)   \01enhance\project1_bay_enhance\bay_data_enhance
RESULTS (Scratch Workspace)   \01enhance\project1_bay_enhance\bay_results_enhance

23

MAKING
SPATIAL
DECISIONS
USING GIS
AND REMOTE
SENSING

2

PROJECT 2
Displaying and enhancing Landsat imagery of Las Vegas, Nevada

Scenario/problem

Visually interpreting Landsat images is important in many qualitative applications. PtD (Protect the Desert) has designated the first segment of the contract to understanding and enhancing grayscale Landsat TM images. PtD has asked for enhanced grayscale imagery that can be used in articles and presentations to highlight different aspects of the city of Las Vegas.

Objectives

You have been asked to use Landsat TM imagery to do the following:
- Display the different bands of Landsat TM data.
- Investigate histograms and their relationship to spectral reflectance.
- Use different enhancement techniques for grayscale imagery to highlight important features of Las Vegas.

DATA (Current Workspace)   \01enhance\project2_vegas_enhance\vegas_data_enhance
RESULTS (Scratch Workspace)   \01enhance\project2_vegas_enhance\vegas_results_enhance

Deliverables

We recommend the following deliverables for this project:

1. A map that identifies Las Vegas, the study area, the rivers, and the highways.
2. A discussion of what bands to use to best show water, roads and highways, and different varieties of vegetation.
3. Histograms comparing reflectance values for bands 1–5 and 7.
4. A document with three different enhanced images with their respective histograms and a written explanation of the enhancement techniques used.

MAKING
SPATIAL
DECISIONS
USING GIS
AND REMOTE
SENSING

2

ENHANCING
IMAGES

The questions for this project are both quantitative and qualitative. They identify key points that should be addressed in your analysis and presentation.

Keeping track of where your data and results are located is always a challenge. In this project, we give the path to access the data (current workspace) and the path to store the results (scratch workspace) in a footnote on each page. The directions will specify whether the results go in the results folder or the results geodatabase.

Examine the data

The data for this project is stored in the folder **\01enhance\project2_vegas_enhance\ vegas_data_enhance**.

View the item description and ArcGIS metadata to investigate the data. The chart helps you organize this information.

Q1 *Investigate the metadata and complete the following chart on your worksheet.*

Layer	Publication information: Who created the data?	Time period data is relevant	Spatial horizontal coordinate system	Data type	Resolution for rasters
Landsat					
Urban					

DATA (Current Workspace)   \01enhance\project2_vegas_enhance\vegas_data_enhance
RESULTS (Scratch Workspace)   \01enhance\project2_vegas_enhance\vegas_results_enhance

25

Organize and document your work

Be sure to refer to project 1 and your process summary.

MAKING
SPATIAL
DECISIONS
USING GIS
AND REMOTE
SENSING

2

ENHANCING
IMAGES

1. Set up the proper directory structure.

2. Create a process summary.

3. Document the map.

4. Set the environments on the Geoprocessing menu:
   a. Data Frame Coordinate System to Projected Coordinate Systems > UTM > WGS 1984 > Northern Hemisphere > WGS 1984 UTM Zone 11N.
   b. Current Workspace to \01enhance\project2_vegas_enhance\vegas_data_enhance.
   c. Scratch Workspace to \01enhance\project2_vegas_enhance\vegas_results_enhance.
   d. Output Coordinate System to Same as Display.

Analysis

An important first step in geospatial analysis is to develop a basemap of your study area.

STEP 1: Create a basemap of Las Vegas, the study area, the rivers, and the highways

Deliverable 1: A map that identifies Las Vegas, the study area, the rivers, and the highways.

Q2 *Write a short description of the geography of Las Vegas and the surrounding area.*

STEP 2: Investigate individual grayscale bands in the Image Analysis window

1. Add L5039035_03520060512_B40 from your data folder and answer the following questions.

Q3 *What features can you identify in this Landsat scene?*

DATA (Current Workspace) \01enhance\project2_vegas_enhance\vegas_data_enhance
RESULTS (Scratch Workspace) \01enhance\project2_vegas_enhance\vegas_results_enhance

Q4   *What is the smallest feature you can identify? Why?*

Remember: You can add an imagery basemap and use the Effects toolbar to swipe or flicker to compare the Landsat imagery to aerial imagery.

2.   Add AOI from your data folder and zoom to AOI 1.

MAKING
SPATIAL
DECISIONS
USING GIS
AND REMOTE
SENSING

2

*ENHANCING
IMAGES*

Q5   *When you select the DRA check box in the Image Analysis window, what happens to the features of the airport?*

3.   Add bands 1–3, 5, and 7 (B10, B20, B30, B50, and B70). Arrange the bands in numerical order.

4.   Zoom to AOI 5.

Q6   *In which band are the streets most visible?*

5.   Zoom to AOI area 4.

Q7   *Which band best delineates water?*

6.   Zoom to AOI area 1.

Q8   *Which band distinguishes different types of vegetation?*

Deliverable 2: A discussion of what bands to use to best show water, roads and highways, and different vegetation varieties.

STEP 3: Match DN to land features

Q9   *Select values representing water in the attribute table, and those values will be highlighted in the image. Using this technique complete the following chart on your worksheet.*

	Range of digital numbers			
	Band 1	Band 2	Band 3	Band 4
water				

MAKING
SPATIAL
DECISIONS
USING GIS
AND REMOTE
SENSING

2

ENHANCING
IMAGES

Q10 *Using the XY icon and the coordinates given below, locate McCarran International Airport. Use the Pixel Inspector to show an array of pixels. Examine the different features of the airport. How do the pixel values change?*

    X: –12819032   Y: 4311863

STEP 4: Create frequency histograms and an all band image frequency histogram

1. Create a histogram graph of L5039035_03520060512_B40.

Q11 *What is the total number of pixels in the image?*

Q12 *How many pixels have a DN of 100?*

Q13 *How many pixels have a DN of approximately 20?*

Q14 *How would you describe the graph?*

Q15 *What does the large number of pixels with a DN of 0 represent?*

2. Make a composite histogram showing all Landsat bands except band 6.

Q16 *Which bands have bimodal histograms?*

Q17 *Which band has the most condensed distribution of reflectance?*

Q18 *Which band has the broadest distribution of reflectance?*

Deliverable 3: Histograms comparing reflectance values for bands 1–5 and 7.

STEP 5: Enhance images

1. Using L5039035_03520060512_B40, answer the following questions.

Q19 *Describe how the pixels are distributed after the standard deviation stretch.*

Q20 *Explain the initial spike on the left side of the graph.*

2. Using clip_band4, answer the following questions.

Q21 *How does clip_band4 look different from band4?*

Q22 *Why does the histogram look different?*

Q23 *How are the pixels distributed when a histogram equalization stretch is used?*

Q24 *Compare the appearance of an image where a histogram equalization stretch has been applied to an image where a standard deviation or minimum-maximum stretch has been applied.*

Q25 *Compare the all band frequency histogram from the Chesapeake Bay to the all band frequency histogram of Las Vegas and the surrounding area. What does the histogram tell you about the land features?*

Deliverable 4: A document with three different enhanced images with their respective histograms and a written explanation of the enhancement techniques used.

MAKING
SPATIAL
DECISIONS
USING GIS
AND REMOTE
SENSING

*ENHANCING
IMAGES*

DATA (Current Workspace)   \01enhance\project2_vegas_enhance\vegas_data_enhance
RESULTS (Scratch Workspace)   \01enhance\project2_vegas_enhance\vegas_results_enhance

29

MAKING
SPATIAL
DECISIONS
USING GIS
AND REMOTE
SENSING

3

PROJECT 3
On your own

Scenario/problem

You have worked through a guided project on Landsat TM satellite imagery for the Chesapeake Bay and repeated the analysis for Las Vegas. In this project, you will reinforce your skills by researching and analyzing a similar scenario entirely on your own. First, you must identify your study area and acquire the data for your analysis. You may want to study a local area. Refer to appendix A for directions on how to download your satellite imagery.

Research

Research the problem and answer the following questions:

1. What is the area of study?
2. What is the problem you are going to study?
3. What data is available?

Obtain the data

Do you have access to baseline data? Data and Maps for ArcGIS at http://www.esri.com/data/data-maps provides many of the layers of data that are needed for project work. Be sure to pay particular attention to the source of data and get the latest version. You can obtain data from the following sources:

- USGS Globalization Viewer at http://glovis.usgs.gov: Access to multiple sets of EROS satellite and aerial imagery.
- Census 2000 TIGER/Line Data at http://www.esri.com/tiger: Access to Census 2000 line data.
- Geospatial One-Stop at http://geo.data.gov: Web-based geospatial resources.
- The National Atlas at http://www.nationalatlas.gov: A range of products and geographic information about the United States.
- The National Map at http://nationalmap.gov/viewer.html: Data includes elevation, land cover, and topographic maps.

MAKING
SPATIAL
DECISIONS
USING GIS
AND REMOTE
SENSING

3

ENHANCING
IMAGES

Workflow

After researching the problem and obtaining the data, you should do the following:

1. Write a brief scenario.

2. State the problem.

3. Define the deliverables.

4. Examine the metadata.

5. Set the directory structure, start your process summary, and document the map.

6. Decide what you need for the data frame coordinate system and the environments.
   a. What is the best projection for your work?
   b. Do you need to set a cell size or mask?

7. Start your analysis.

8. Prepare your presentation and deliverables.

9. Always remember to document your work in a process summary.

MODULE 2
COMPOSITE IMAGES

Introduction

Satellite-based cameras and instruments can produce very high-resolution imagery. However, each image is typically captured at a particular wavelength. So when you combine images of the same scene at a number of wavelengths, the whole becomes much greater than the sum of its parts. In this module, you will produce composite images in both true and false color and explore how combining different wavelengths allows you to differentiate the features in the scene.

Scenarios in this module

- Creating multispectral imagery of the Chesapeake Bay
- Creating multispectral imagery of Las Vegas, Nevada
- On your own

Student worksheets

The student worksheet files can be found on the Maps and Data DVD.

Project 1: Chesapeake Bay student sheet
- File name: 02a_composite_worksheet.docx
- Location: \Student_Worksheets\02composite

Project 2: Las Vegas student sheet
- File name: 02b_composite_worksheet.docx
- Location: \Student_Worksheets\02composite

MAKING

SPATIAL

DECISIONS

USING GIS

AND REMOTE

SENSING

PROJECT 1
Creating multispectral imagery of the Chesapeake Bay

Background

Remote sensing images from instruments such as Landsat are typically grayscale and taken through a particular diffraction grating or filter. However, images of the same area taken with different filters can be combined to make a color scene. A color scene (image) is generated by mixing three grayscale images representing blue, green, and red. Imagine needing three channels or three color guns (blue, green, and red) to create a color image. If each of these grayscale images actually corresponds to these colors, a natural color (or true color) image can be produced. In ArcGIS, Landsat satellite imagery can be added by dragging the metadata of the Landsat scene to the ArcMap Table of Contents. A default RGB band combination is produced, and the bands are named appropriately. Band combinations can be produced that highlight or display land and water features differently. The following table describes the spectral range of each band. Keep in mind that band combinations will have different interpretations for different sensors.

DATA (Current Workspace) \02composite\project1_bay_comp\bay_data_comp
RESULTS (Scratch Workspace) \02composite\project1_bay_comp\bay_results_comp

DESCRIPTIONS OF LANDSAT SPECTRAL BANDS

Band	Wavelength	Micrometers	ArcGIS name
1	blue	0.45–0.52	Blue
2	green	0.52–0.60	Green
3	red	0.63–0.69	Red
4	near infrared	0.76–0.90	NearInfrared_1
5	shortwave infrared 1	1.55–1.75	NearInfrared_2
6	thermal	10.40–12.50	
7	shortwave infrared 2	2.08–2.35	MidInfrared

MAKING
SPATIAL
DECISIONS
USING GIS
AND REMOTE
SENSING

1

COMPOSITE
IMAGES

The following list includes a short summary of the characteristics of the most common band combinations for Landsat TM imagery. The combination is listed based on which bands would go into the red, green, and blue channels, respectively. For example, for the 752 combination, band 7 would be displayed in the red channel, band 5 in the green, and band 2 in the blue.

- 321: This band combination creates a true color or natural-looking image. This band combination is useful for bathymetric and coastal studies.
- 432: Using band 4 in the red channel results in more sharply defined water boundaries than in the 321 image. By displaying the band that senses peak chlorophyll reflectance (band 4) as red, vegetation appears red. Generally, deep-red hues indicate broad leaf and/or healthier vegetation, and lighter reds signify grasslands or sparsely vegetated areas. Densely populated urban areas appear as light blue.
- 742: This combination retains the benefits of using the infrared bands yet presents vegetation in familiar green tones. Shortwave infrared band 7 helps discriminate moisture content in both vegetation and soils. Urban areas appear in varying shades of magenta. Grasslands appear as light green.
- 453: With this band combination, vegetation type and condition are displayed as variations of hue (browns, greens, and oranges). This band combination highlights moisture differences and is useful in analysis of soil and vegetation conditions. Generally, the wetter the soil, the darker it appears.

Scenario/problem

Chesapeake Bay Foundation managers are asking for an analysis of multiband spectral imagery of the Chesapeake Bay for specific areas of interest. They want to use composite images to monitor urban growth, study the extent and patterns of turbidity, observe beach erosion, measure shoreline change, and map water pollution, among other things. Your job is to generate color composite images and identify areas of interest to the CBF.

DATA (Current Workspace)   \02composite\project1_bay_comp\bay_data_comp
RESULTS (Scratch Workspace)   \02composite\project1_bay_comp\bay_results_comp

35

Objectives

For this project, you will use Landsat imagery to do the following:

- Produce color composite images.
- Use color composite images to observe land features.
- Compare and contrast land features using different color composite images.
- Create data driven pages for Landsat scene investigation.

MAKING
SPATIAL
DECISIONS
USING GIS
AND REMOTE
SENSING

1

COMPOSITE
IMAGES

Deliverables

We recommend the following deliverables for this exercise:

1. A map with multiple data frames showing different band combinations focused on different areas of interest.
2. A written assessment of the information offered by the different band combinations.
3. A layout with data driven pages and an index layer that divides the map into sections based on each index feature.

The questions for this project are both quantitative and qualitative. They identify key points that should be addressed in your analysis and presentation.

Keeping track of where your data and results are located is always a challenge. In these projects, we give the path to access the data (current workspace) and the path to store the results (scratch workspace) in a footnote on each page. The directions will specify whether the results go in the results folder or the results geodatabase.

Examine the data

This section was completed in module 1.

Organize and document your work

The following preliminary steps are essential to a successful geospatial analysis.

Examine the directory structure

In a geospatial project, you must carefully keep track of the data and your calculations. You will work with a number of different files, and it is important to keep them organized so you can easily find them. The best way to do this is to have a folder for your project that contains a data folder. For this project, the folder named **\02composite\project1_bay_comp\bay_data_comp** will be your project folder. Make sure it is stored in a place where you have write access.

DATA (Current Workspace)   \02composite\project1_bay_comp\bay_data_comp
RESULTS (Scratch Workspace)   \02composite\project1_bay_comp\bay_results_comp

You can store your data inside the results folder. The results folder already contains an empty geodatabase named **bay_results** for this purpose. Save your map documents to the **bay_results_comp** folder.

MAKING
SPATIAL
DECISIONS
USING GIS
AND REMOTE
SENSING

1

COMPOSITE
IMAGES

Folder structure:
02composite
       project1_bay_comp
          bay_data_comp
             bay.gdb
             landsat_may_2006
          bay_ results_comp
             bay_results.gdb

Create a process summary

The process summary is simply a list of the steps you used to do your analysis. We suggest using a simple text document for your process summary. Keep adding to it as you do your work to avoid forgetting any steps. The following list shows an example of the first few entries in a process summary:

1. Prepare composite bands.
2. Produce a map with multiple data frames showing different band combinations focused on different areas of interest.

A process summary is essentially a work log. It will allow you to remember the steps in your data analysis at a later time or communicate them with others who might need to reproduce your work.

Document the map

1. Start ArcMap and add descriptive properties to your map document properties.

2. Be sure to select the pathnames check box to store relative pathnames to all your data.

Set the environments

1. On the View menu, click Data Frame Properties. Set the map projection to Projected Coordinate Systems > UTM > WGS 1984 > Northern Hemisphere > WGS 1984 UTM Zone 18N.

2. Set the Current Workspace to \02composite\project1_bay_comp\bay_data_comp.

DATA (Current Workspace) \02composite\project1_bay_comp\bay_data_comp
RESULTS (Scratch Workspace) \02composite\project1_bay_comp\bay_results_comp

3. Set the Scratch Workspace to \02composite\project1_bay_comp\bay_results_comp\bay_ results.gdb.

4. For Output Coordinate System, select Same as Display.

MAKING
SPATIAL
DECISIONS
USING GIS
AND REMOTE
SENSING

Analysis

Once you have examined the data, completed the map documentation, and set the environments, you are ready to begin the analysis and complete the data displays you need to address the problem. For this module, you have been asked to create different color composite images.

STEP 1: Create different color composite images

1. Name the data frame RGB_321.

In the Catalog window in the landsat_may_2006 folder, the Landsat bands are presented as individual TIFF images, and in module 1 you worked with the bands as individual images. In this module, you will work with raster products that combine the bands. The L5015033_ 03320060504_MTL file is the metadata that is produced when a scene is processed through the Level 1 Product Generation System. If you click the MTL file, it will expand to show the derived raster products. In this case, you will see Multispectral.

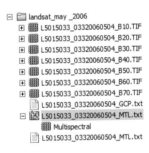

2. Drag the metadata file L5015033_03320060504_MTL to the ArcMap Table of Contents.

3. For the Multispectral layer, go to Properties > Symbology. When asked if you want to estimate histograms, click Yes.

4. When the Symbology dialog box appears, reselect the bands by pulling down the Band menu and clicking Red, Green, and Blue, respectively. Now the image is displayed in natural color, and names indicating the wavelengths of the different bands are shown.

```
□ ⊜ RGB_321
    □ ☑ Multispectral_L5015033_03320060504_MTL
          RGB
        ■ Red:   Red
        ■ Green: Green
        ■ Blue:  Blue
```

MAKING
SPATIAL
DECISIONS
USING GIS
AND REMOTE
SENSING

When this composite image was created, ArcMap applied several functions on the fly as the data was accessed and then viewed. These functions can be examined from the Properties menu.

5. Click the Multispectral file and go to Properties > Functions to view the functions applied.

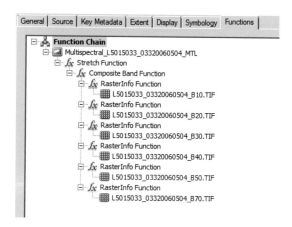

Q1 *List and describe the two functions that have been applied to the raster data.*

Q2 *Why is this image referred to as a natural color image?*

Q3 *Describe three features and three types of land cover in this natural color image.*

Q4 *When looking at the whole Landsat scene, you will see a band of gray going through the middle of the image from the southwest to the northeast. What type of land cover does this represent? Does the gray represent urban land? Agriculture? Zoom in closer to answer.*

Q5 *Describe any variations you see in the bodies of water.*

DATA (Current Workspace)   \02composite\project1_bay_comp\bay_data_comp
RESULTS (Scratch Workspace)   \02composite\project1_bay_comp\bay_results_comp

39

These color composite images are easier to analyze if you zoom to an interesting area. The following image is zoomed to the Washington, DC, area. The band combination for the image is also shown, indicating that it is a true color (321) image.

MAKING
SPATIAL
DECISIONS
USING GIS
AND REMOTE
SENSING

COMPOSITE
IMAGES

6. Add states to the Table of Contents and make them hollow with a yellow outline.

7. Copy the data frame, and on the main menu pull down the Edit menu and click Paste. Change the data frame name to RGB_432. Right-click and activate this data frame.

8. Zoom out to the entire Landsat scene.

DATA (Current Workspace) \02composite\project1_bay_comp\bay_data_comp
RESULTS (Scratch Workspace) \02composite\project1_bay_comp\bay_results_comp

9. Go to Properties > Symbology and change the red channel to NearInfrared_1, the green channel to Blue, and the blue channel to Green.

MAKING
SPATIAL
DECISIONS
USING GIS
AND REMOTE
SENSING

1

COMPOSITE
IMAGES

Q6 *Which wavelengths are now being displayed by each of the three RGB channels?*

Q7 *Why is it important to know which color is assigned to which band?*

Q8 *What color represents vegetation in an RGB 432 image?*

Q9 *What color represents developed land in an RGB 432 image?*

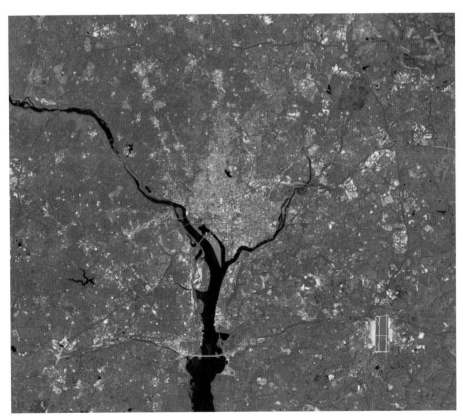

DATA (Current Workspace)   \02composite\project1_bay_comp\bay_data_comp
RESULTS (Scratch Workspace)   \02composite\project1_bay_comp\bay_results_comp

41

MAKING
SPATIAL
DECISIONS
USING GIS
AND REMOTE
SENSING

1

COMPOSITE
IMAGES

10. Repeat directions 7–9. Name the data frame RGB_742. Go to Properties > Symbology and change the red channel to MidInfrared, the green channel to NearInfrared_1, and the blue channel to Green.

Q10 **Which wavelengths of light are now being displayed by each of the three RGB channels?**

Q11 **What colors represent vegetation in an RGB 742 image?**

Q12 **What color represents developed land in an RGB 742 image?**

11. Repeat directions 7–9. Name the data frame RGB_453. Go to Properties > Symbology and change the red channel to NearInfrared_1, the green channel to NearInfrared_2, and the blue channel to Red.

Q13 **Which wavelengths of light are now being displayed by each of the three RGB channels?**

Q14 **What colors represent vegetation in an RGB 453 image?**

Q15 **Is vegetation represented in different shades in the RGB 742 color composite? Why?**

Q16 **What colors represent developed land in an RGB 453 image?**

12. Save the map document as Comp_Bay1.

13. Save the map document a second time as Comp_Bay2.

Deliverable 1: A map with multiple data frames showing different band combinations focused on different areas of interest.

DATA (Current Workspace)   \02composite\project1_bay_comp\bay_data_comp
RESULTS (Scratch Workspace)   \02composite\project1_bay_comp\bay_results_comp

STEP 2: Analyze color composite images

You are now ready to do a qualitative visual interpretation of the images. You were initially asked to do the following using Landsat imagery:

- Produce color composite images.
- Use color composite images to observe land features.
- Compare and contrast land features using different color composite images.
- Create data driven pages for Landsat scene investigation.

You have produced the color composite images, and now you are ready to observe, compare, and contrast land features using different color composite images.

MAKING
SPATIAL
DECISIONS
USING GIS
AND REMOTE
SENSING

1

COMPOSITE
IMAGES

1. Open the map document Comp_Bay2.

2. Remove all the data frames except RGB_321. Rename the RGB_321 data frame Composites. Zoom to the full extent.

3. Change the name of Multispectral_ L5015033_03320060504_MTL to RGB_321.

4. Copy and paste RGB_321. Rename it RGB_432 and adjust the displayed bands accordingly.

5. Repeat directions 3 and 4 and produce RGB_742 and RGB_453.

6. Add the feature class AOI from the bay geodatabase.

The area of interest (AOI) feature class helps you focus on different areas of interest.

Your Table of Contents window should be arranged like the following screen capture.

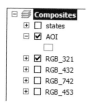

Arranging your raster features in this way allows you to compare the different composites by turning them on and off.

DATA (Current Workspace)   \02composite\project1_bay_comp\bay_data_comp
RESULTS (Scratch Workspace)   \02composite\project1_bay_comp\bay_results_comp

43

7. Make the AOI feature class hollow and label it. You might want to increase the size of the labels and make them a different color. Focus on these areas in your work. Use the following questions to help in your analysis.

MAKING
SPATIAL
DECISIONS
USING GIS
AND REMOTE
SENSING

1

COMPOSITE
IMAGES

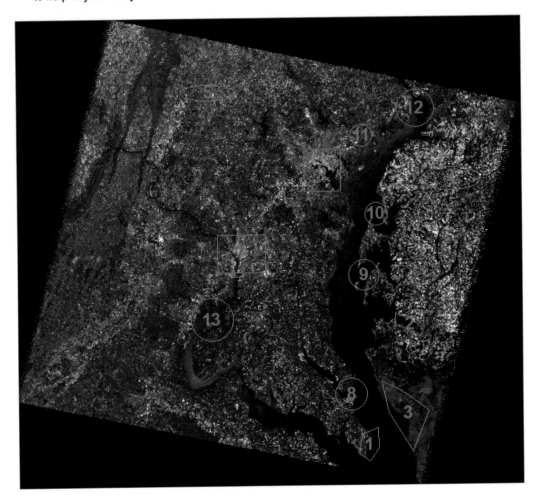

Q17 *Where are the largest concentrations of urban population?*

Q18 *Discuss where the urban population will most likely expand.*

Q19 *From investigating the Landsat scene, can you suggest areas that are vulnerable to natural disasters?*

Q20 *From investigating the Landsat scene, can you suggest areas that could be targeted for restoration?*

8. Save your map document.

Deliverable 2: A written assessment of the different band combinations.

DATA (Current Workspace)   \02composite\project1_bay_comp\bay_data_comp
RESULTS (Scratch Workspace)   \02composite\project1_bay_comp\bay_results_comp

STEP 3: Create data driven pages and an index layer

Data driven pages allow you to quickly and easily create a series of layout pages from a single map document (such as you might use to create a map book). An index layer divides the map into sections based on an index feature in the layer and generates one page per index feature. In this instance, you want to access different sections of the image for analysis not only in that section, but in all the different band composites. The following image shows an example of an index layer that would allow you to look at one page (section) at a time. You could look at page A1 or page D5, depending on how you navigated through the pages.

MAKING
SPATIAL
DECISIONS
USING GIS
AND REMOTE
SENSING

1

COMPOSITE
IMAGES

A1	A2	A3	A4	A5
B1	B2	B3	B4	B5
C1	C2	C3	C4	C5
D1	D2	D3	D4	D5

1. Start ArcMap and add descriptive properties to your map document properties.

2. Be sure to select the pathnames check box to store relative pathnames to all your data.

3. On the View menu, click Data Frame Properties. Set the map projection to Projected Coordinate Systems > UTM > WGS 1984 > Northern Hemisphere > WGS 1984 UTM Zone18N.

4. Set the Current Workspace to \02composite\project1_bay_comp\bay_data_comp.

5. Set the Scratch Workspace to \02composite\project1_bay_comp\bay_results_comp\bay_results.gdb.

DATA (Current Workspace)   \02composite\project1_bay_comp\bay_data_comp
RESULTS (Scratch Workspace)   \02composite\project1_bay_comp\bay_results_comp

MAKING
SPATIAL
DECISIONS
USING GIS
AND REMOTE
SENSING

1

COMPOSITE
IMAGES

6. For Output Coordinate System, select Same as Display.

7. Add dd_studyarea and L5015033_03320060504_MTL.

8. Go to Properties > Symbology. When asked if you want to estimate histograms, click Yes.

9. When the Symbology dialog box appears, reselect the bands by clicking Red, Green, and Blue, as you did earlier. Name the layer RGB_321.

10. Copy RGB_321 and paste to create a second layer. Change the red channel to NearInfrared_1, the green channel to Red, and the blue channel to Green. Name the layer RGB_432.

11. Repeat direction 10. Change the red channel to MidInfrared, the green channel to NearInfrared_1, and the blue channel to Green. Name the layer RGB_742.

12. Repeat direction 10. Change the red channel to NearInfrared_1, the green channel to NearInfrared_2, and the blue channel to Red. Name the file RGB_453.

13. Set the scale to 1:150,000.

Grid index features can be used to define the spatial extent of your map as well as label the pages. A grid index feature allows you to divide your map into a series of pages.

14. Search for the tool "grid index features." Run the Grid Index Features tool and use the following parameters. This process will create a new feature class called index_150.
    a. Output Feature Class is index_150 in the results folder.
    b. Input Feature is dd_studyarea.
    c. Select Use Page Unit and Scale.
    d. Set the map scale to 1:150,000.
    e. Set the polygon width to 7.5 inches.
    f. Set the height to 9 inches.
    g. Click OK.

Index_150 is a feature class that is also a grid index. It has six columns and four rows and is the same size as the dd_studyarea.

15. Remove dd_studyarea.

16. Switch to Layout View.

17. Turn on the Data Driven Pages toolbar.

MAKING
SPATIAL
DECISIONS
USING GIS
AND REMOTE
SENSING

18. Using the Data Driven Pages toolbar, do the following:
  a. Click the Data Driven Page Setup icon and select the Enable Data Driven Pages check box.
  b. Click the Extent tab, and then click Center and Maintain Current Scale.
  c. In the Table of Contents, make index_150 hollow, and make the outline red with a width of 2.
  d. Label the index_150 layer by page name. Make the labels red also.

19. Zoom to page A1 of the index.

You now have control of your study area in two ways. In the Table of Contents, you can turn on and off the composite you want to see. For example, you can select RGB_321 or RGB_432. You can also move around on your map from page to page by typing the page number you want to access or clicking the arrow on the Data Driven Pages toolbar.

Or you can create a data frame index that allows you to navigate from frame to frame.

20. Insert a new data frame and name it Index.

21. Copy the index_150 layer and paste it to the Index data frame.

22. In Layout View, relocate this data frame to the bottom section of the map layout.

23. Click the feature class Index_150 and go to Properties > Definition Query.

24. Click Page Definition and then Enable.

25. Click OK and then Apply.

DATA (Current Workspace)   \02composite\project1_bay_comp\bay_data_comp
RESULTS (Scratch Workspace)   \02composite\project1_bay_comp\bay_results_comp

47

You can now analyze specific areas of the image. You can access the pages by either typing their number or opening the attribute table from the Data Driven Pages toolbar and selecting them there.

MAKING
SPATIAL
DECISIONS
USING GIS
AND REMOTE
SENSING

1

COMPOSITE
IMAGES

26. Save your map document as Comp_Bay3.

Q21 *Examine pages A2, B2, C4, and D2. Write a detailed analysis of these areas. Remember to look at all four band composites.*

Deliverable 3: A layout with data driven pages and an index layer that divides the map into sections based on each index feature.

Once your analysis is complete, you will still need to develop a solution to the original problem and present your results in a compelling way for this particular situation. The presentation of your various data displays must explain what they show and how they help solve the problem identified by the Chesapeake Bay Foundation.

DATA (Current Workspace)   \02composite\project1_bay_comp\bay_data_comp
RESULTS (Scratch Workspace)   \02composite\project1_bay_comp\bay_results_comp

MAKING
SPATIAL
DECISIONS
USING GIS
AND REMOTE
SENSING

PROJECT 2
Creating multispectral imagery of Las Vegas, Nevada

Scenario/problem

The urban planners of Las Vegas are asking PtD for a multispectral band analysis of Las Vegas and the surrounding area. They want to use color composite images to monitor patterns of urban growth, look at the increase of impervious areas, and gather information about the encroachment of the urban area on the desert environment.

Objectives

You will use Landsat TM imagery to do the following:
- Produce color composite images.
- Use color composite images to observe land features.
- Compare and contrast land features using different color composite images.
- Create data driven pages for Landsat scene investigation.

Deliverables

We recommend the following deliverables for this project:
1. A map with multiple data frames showing different band combinations focused on different areas of interest.
2. A written assessment of the information offered by the different band combinations.

DATA (Current Workspace)   \02composite\project2_vegas_comp\vegas_data_comp
RESULTS (Scratch Workspace)   \02composite\project2_vegas_comp\vegas_results_comp

49

MAKING
SPATIAL
DECISIONS
USING GIS
AND REMOTE
SENSING

2

COMPOSITE
IMAGES

3. A layout with data driven pages with an index layer that divides the map into sections based on each index feature.

The questions for this project are both quantitative and qualitative. They identify key points that should be addressed in your analysis and presentation.

Keeping track of where your data and results are located is always a challenge. In these projects, we give the path to access the data (current workspace) and the path to store the results (scratch workspace) in a footnote on each page. The directions will specify whether the results go in the results folder or the results geodatabase.

Examine the data

This section was completed in module 1, project 2.

Organize and document your work

The data for this project is stored in the **\02composite\project2_vegas_comp\vegas_data_ comp** folder. Be sure to refer to project 1 and your process summary.

1. Set up the proper directory structure.

2. Create a process summary.

3. Document the map.

4. Set the environments on the Geoprocessing menu:
   a. Data frame coordinate system to Projected Coordinate Systems > UTM > WGS 1984 > Northern Hemisphere > WGS 1984 UTM Zone18N.
   b. Current Workspace to \02composite\project2_vegas_comp\vegas_data_comp.
   c. Scratch Workspace to \02composite\project2_vegas_comp\vegas_results_comp.
   d. Output Coordinate System to Same as Display.

Analysis

Use the skills you acquired in project 1 to help you create true and false color composite images of Las Vegas, Nevada.

STEP 1: Create different color composite images

1. Create color composite images of Las Vegas, Nevada.

Q1   *List and describe the two functions that have been applied to the raster data.*

Q2   *Why is this image referred to as a natural color image?*

Q3   *Describe three features and three types of land cover in this natural color image.*

Q4   *How is this Landsat scene different from the Chesapeake Bay scene you used in project 1?*

Q5   *Zoom in and examine Lake Mead. Lake Mead is the large body of water located in the eastern section of the image. Describe any variations you see in the water.*

Q6   *When looking at the RGB_432 composite, you see concentrated areas of bright red. Zoom in on these areas. What are these areas? What type of vegetation does this represent?*

Q7   *Which false color composite shows Hoover Dam the most clearly? Hoover Dam is located at the southwestern arm of Lake Mead. It is east of Las Vegas and located on the Colorado River. It bridges the states of Nevada and Arizona. What about the islands in the reservoir?*

Deliverable 1: A map of multiple data frames showing different band combinations focused on different areas of interest.

STEP 2: Analyze color composite images

1. Add the AOI and dtl_wat feature classes from your data folder.

Q8   *Does the water body in AOI 3 look like the other water bodies you have observed? Does it look like Lake Mead? Can you suggest why it looks different?*

MAKING
SPATIAL
DECISIONS
USING GIS
AND REMOTE
SENSING

2

COMPOSITE
IMAGES

MAKING
SPATIAL
DECISIONS
USING GIS
AND REMOTE
SENSING

2

COMPOSITE
IMAGES

Q9 *Compare and contrast the land features and the vegetation of AOI 1 and AOI 2.*

Q10 *Where is the concentration of urban population?*

Q11 *Looking at the land features, in what direction do you think the urban area will expand?*

Deliverable 2: A written assessment of the different band combinations.

STEP 3: Create data driven pages and an index layer

Deliverable 3: A layout with data driven pages and an index layer that divides the map into sections based on each index feature.

Q12 *Examine pages B1, B4, C5, D1, and D5. Write a detailed analysis of these areas. Remember to look at all four band combinations.*

DATA (Current Workspace)   \02composite\project2_vegas_comp\vegas_data_comp
RESULTS (Scratch Workspace)   \02composite\project2_vegas_comp\vegas_results_comp

MAKING
SPATIAL
DECISIONS
USING GIS
AND REMOTE
SENSING

3

*COMPOSITE
IMAGES*

PROJECT 3
On your own

Scenario/problem

You have worked through a project on Landsat multispectral imagery for the Chesapeake Bay and repeated the analysis for Las Vegas. In this project, you will reinforce your skills by researching and analyzing a similar scenario entirely on your own. First, you must identify your study area and acquire the data for your analysis. You may want to study a local area. Refer to appendix A for directions on how to download your satellite imagery.

Research

Research the problem and answer the following questions:
1. What is the area of study?
2. What is the problem you are going to study?
3. What data is available?

Obtain the data

Do you have access to baseline data? Data and Maps for ArcGIS at http://www.esri.com/data/data-maps provides many of the layers of data that are needed for project work. Be sure to pay particular attention to the source of data and get the latest version. You can obtain data from the following sources:
*   USGS Globalization Viewer at http://glovis.usgs.gov: Access to multiple sets of EROS satellite and aerial imagery.

MAKING
SPATIAL
DECISIONS
USING GIS
AND REMOTE
SENSING

3

COMPOSITE
IMAGES

- Census 2000 TIGER/Line Data at http://www.esri.com/tiger: Access to Census 2000 line data.
- Geospatial One-Stop at http://geo.data.gov: Web-based geospatial resources.
- The National Atlas at http://www.nationalatlas.gov: A range of products and geographic information about the United States.
- The National Map at http://nationalmap.gov/viewer.html: Data includes elevation, land cover, and topographic maps.

Workflow

After researching the problem and obtaining the data, you should do the following:

1. Write a brief scenario.

2. State the problem.

3. Define the deliverables.

4. Examine the metadata.

5. Set the directory structure, start your process summary, and document the map.

6. Decide what you need for the data frame coordinate system and the environments.
   a. What is the best projection for your work?
   b. Do you need to set a cell size or mask?

7. Start your analysis.

8. Prepare your presentation and deliverables.

9. Always remember to document your work in a process summary.

MODULE 3
SPECTRAL SIGNATURES

Introduction

Scientists can use remote sensing images to identify different features in a scene without ever actually being there. However, to do this, they need to know how different objects (e.g., trees, roads, sidewalks) reflect and absorb light. The absorption characteristics are called spectral signatures. In this module, you will learn how to create and use spectral signatures to analyze imagery.

Scenarios in this module

- Investigating spectral signatures of the Chesapeake Bay
- Investigating spectral signatures of Las Vegas, Nevada
- On your own

Student worksheets

The student worksheet files can be found on the Maps and Data DVD.

Project 1: Chesapeake Bay student sheet
- File name: 03a_signatures_worksheet
- Location: \Student_Worksheets\03signatures

Project 2: Las Vegas student sheet
- File name: 03b_signatures_worksheet
- Location: \Student_Worksheets\03signatures

MAKING
SPATIAL
DECISIONS
USING GIS
AND REMOTE
SENSING

PROJECT 1
Investigating spectral signatures of the Chesapeake Bay

Background

Spectral signatures derived from multispectral images have been in regular use since the 1970s. Today, with the growing importance of hyperspectral imaging (viewing scenes using images taken at a variety of different wavelengths), it is even more important to understand the basics of spectral signatures. Spectral signatures are essentially plots of the reflected radiation of different objects collected using the different wavelength filters of the satellite sensor.

Objects behave differently with different wavelengths of incident energy and, therefore, have unique spectral signatures. Thus, using spectral signatures is one of the important tools in the identification of and discrimination between various objects in analysis of digital data (Kachhwaha 1983, 685). Spectral signatures often can be used to distinguish man-made materials, different types of vegetation and their condition, and other spectrally similar materials.

Scenario/problem

The Chesapeake Bay Foundation knows that chemical spills have historically threatened the Chesapeake Bay. The fragile resources of the coastal region of the bay are particularly vulnerable to harmful spills of oil or other toxins. CBF has heard that spectral signatures can be used to determine the effects of these types of hazards. However, the foundation needs to be educated

DATA (Current Workspace)   \03signatures\project1_bay_ss\bay_data_ss
RESULTS (Scratch Workspace)   \03signatures\project1_bay_ss\bay_results_ss

on spectral signatures before it invests any resources in this application. CBF has asked you to provide a briefing explaining the use of spectral signatures to examine the bay.

Objectives

For this project, you will do the following:

MAKING
SPATIAL
DECISIONS
USING GIS
AND REMOTE
SENSING

- Describe and understand the data collection needed to construct spectral signatures.
- Graph spectral signatures of different features.
- Analyze how feature properties can affect spectral signatures by
  - comparing spectral signatures of different types of vegetation, and
  - comparing spectral signatures of clear and turbid water.

Deliverables

We recommend the following deliverables for this project:

1. A chart and graph of spectral signatures of concrete, grass, and water.
2. A chart and graph comparing spectral signatures of different types of vegetation.
3. A chart and graph comparing spectral signatures of clear and turbid water.

The questions for this project are both quantitative and qualitative. They identify key points that should be addressed in your analysis and presentation.

Keeping track of where your data and results are located is always a challenge. In these projects, we give the path to access the data (current workspace) and the path to store the results (scratch workspace) in a footnote on each page. The directions will specify whether the results go in the results folder or the results geodatabase.

Examine the data

This section was completed in module 1.

Organize and document your work

The following preliminary steps are essential to a successful geospatial analysis.

Examine the directory structure

In a geospatial project, you must carefully keep track of the data and your calculations. You will work with a number of different files, and it is important to keep them organized so you can easily find them. The best way to do this is to have a folder for your project that contains a data

DATA (Current Workspace)   \03signatures\project1_bay_ss\bay_data_ss
RESULTS (Scratch Workspace)   \03signatures\project1_bay_ss\bay_results_ss

59

MAKING
SPATIAL
DECISIONS
USING GIS
AND REMOTE
SENSING

1

SPECTRAL
SIGNATURES

folder. For this project, the folder named **\03signatures\project1_bay_ss\bay_data_ss** will be your project folder. Make sure it is stored in a place where you have write access.

You can store your data inside the results folder. The results folder already contains an empty geodatabase named **bay_results** for this purpose. Save your map documents inside the **bay_results_ss** folder.

Folder structure:
>03signatures
>>project1_bay_ss
>>>bay_data_ss
>>>>bay.gdb
>>>>landsat_may_2006
>>>bay_results_ss
>>>>bay_results.gdb

Create a process summary

The process summary is simply a list of the steps you used to do your analysis. We suggest using a simple text document for your process summary. Keep adding to it as you do your work to avoid forgetting any steps. The following list shows an example of the first few entries in a process summary:

1. Add the signature points.
2. Complete a chart of DNs of signature points.
3. Produce graphs of concrete, grass, and water.

A process summary is essentially a work log. It will allow you to remember the steps in your data analysis at a later time or communicate them with others who might need to reproduce your work.

Document the map

1. Start ArcMap and add descriptive properties to your map document.

2. Be sure to select the pathnames check box to store relative pathnames to all your data.

Set the environments

1. On the View menu, click Data Frame Properties. Set the map projection to Projected Coordinate Systems > UTM > WGS 1984 > Northern Hemisphere > WGS 1984 UTM Zone 18N.

DATA (Current Workspace)   \03signatures\project1_bay_ss\bay_data_ss
RESULTS (Scratch Workspace)   \03signatures\project1_bay_ss\bay_results_ss

2. Set the Current Workspace to \03signatures\project1_bay_ss\bay_data_ss.

3. Set the Scratch Workspace to \03signatures/project1_bay_ss\bay_results_ss.

4. For Output Coordinate System, select Same as Display.

Analysis

Once you have examined the data, completed the map documentation, and set the environments, you are ready to begin the analysis and complete the data displays you need to address the problem. For this module, you have been asked to provide an explanation of spectral signatures.

STEP 1: Create three standard spectral signatures

1. Add the L5015033_03320060504_MTL file from the landsat_map_2006 data folder.

2. Go to Properties > Symbology and reselect the band colors to make the RGB composite (red, green, and blue). Click Apply. This produces a natural color image.

3. Add the points concrete, grass, and water from the ss_points feature dataset in your data folder. Make the points different colors.

4. Add an online basemap, such as Imagery.

Remember: You can add different basemaps using the Add Data button on the Standard toolbar. There are a variety of basemaps to choose from. These basemaps will provide imagery, transportation, and boundaries and places. It will help you identify where the points are located. Basemaps such as Imagery provide high-resolution color views of a scene. They are useful for identifying specific landmarks on the ground, but they do not contain usable multispectral data. The combination of these basemaps and Landsat TM data aids analysis.

MAKING
SPATIAL
DECISIONS
USING GIS
AND REMOTE
SENSING

1

SPECTRAL
SIGNATURES

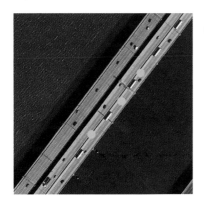

Q1   ***Where are the concrete points located?***

DATA (Current Workspace)   \03signatures\project1_bay_ss\bay_data_ss
RESULTS (Scratch Workspace)   \03signatures\project1_bay_ss\bay_results_ss

61

MAKING
SPATIAL
DECISIONS
USING GIS
AND REMOTE
SENSING

1

SPECTRAL
SIGNATURES

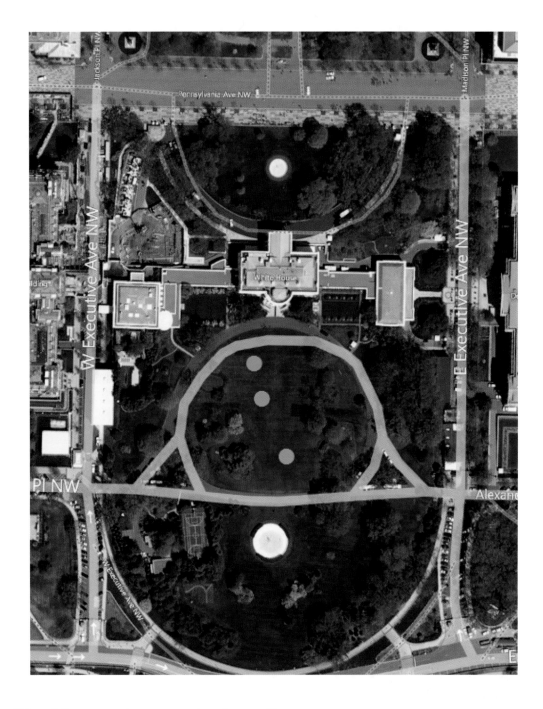

Q2 *Where are the grass points located?*

DATA (Current Workspace)   \03signatures\project1_bay_ss\bay_data_ss
RESULTS (Scratch Workspace)   \03signatures\project1_bay_ss\bay_results_ss

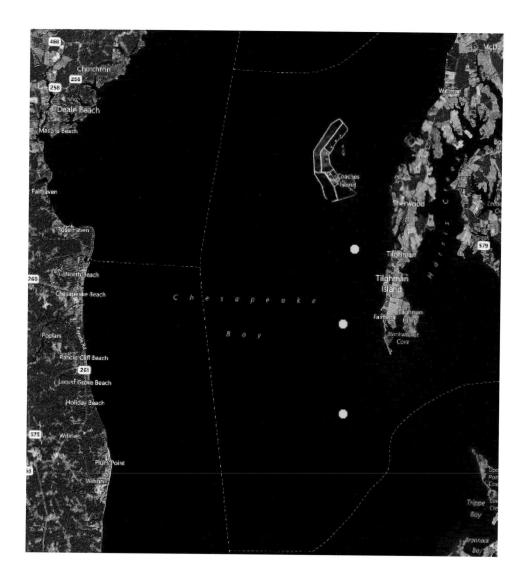

MAKING
SPATIAL
DECISIONS
USING GIS
AND REMOTE
SENSING

1

SPECTRAL
SIGNATURES

Q3 *Where are the water points located?*

These three features—concrete, grass, and water—have distinctive spectral response patterns. Finding distinctive spectral response patterns is the key to interpreting remotely sensed imagery using spectral signatures. The spectral signature of vegetation changes with the time of year, and, as you will see, the spectral signature of water depends on the clarity and constituents/contaminants in the water, such as algae or nitrates.

In order to interpret the spectral signatures, the digital numbers of each of the features must be collected, tabulated, and plotted.

DATA (Current Workspace)   \03signatures\project1_bay_ss\bay_data_ss
RESULTS (Scratch Workspace) \03signatures\project1_bay_ss\bay_results_ss

63

MAKING
SPATIAL
DECISIONS
USING GIS
AND REMOTE
SENSING

1

*SPECTRAL
SIGNATURES*

5. Remove the online resources.

6. Tabulate and plot the concrete points.
   a. Zoom to the concrete points.

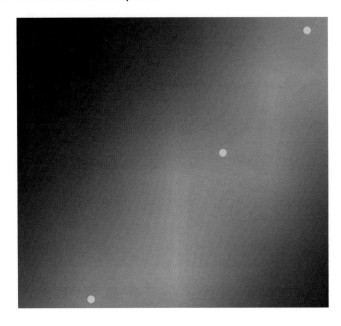

   b. Using the Identify tool, click each of the points and record its digital number.
   c. When using the Identify tool, select Multispectral_L5015033_03320060504_MTL on the tab that says Identify from.
   d. The Identify tool will give you the digital number of each of the three bands in the composite. In this case, you will see the digital numbers for red, green, and blue, because the Landsat scene is displayed in natural color.

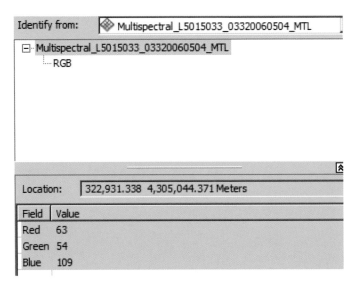

DATA (Current Workspace)   \03signatures\project1_bay_ss\bay_data_ss
RESULTS (Scratch Workspace)   \03signatures\project1_bay_ss\bay_results_ss

e. To get the digital numbers for NearInfrared_1, NearInfrared_2, and MidInfrared (shortwave infrared 2), go to Properties > Symbology and change the channel combinations as shown in the following screen capture.

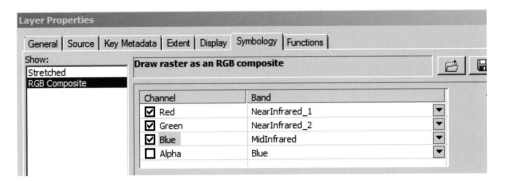

MAKING
SPATIAL
DECISIONS
USING GIS
AND REMOTE
SENSING

1

SPECTRAL
SIGNATURES

f. Record the digital number of each of the points for each of the bands in the following chart.

Band #	Band Name	Grass			Average	Water			Average	Concrete			Average
B10	Blue												
B20	Green												
B30	Red												
B40	NearInfrared_1												
B50	NearInfrared_2												
B70	MidInfrared												

g. Average the three readings.

7. Repeat direction 6 for water and grass.

8. Make a spreadsheet or a comma-delimited text file (.csv) to store your data and bring it into ArcMap. The spreadsheet or comma-delimited text file should have the information shown in the chart.

Band #	Band Name	Grass	Water	Concrete
B10	Blue			
B20	Green			
B30	Red			
B40	NearInfrared_1			
B50	NearInfrared_2			
B70	MidInfrared			

9. Save the spreadsheet as ss_bay to your results folder.

The spectral signature of a particular material can be viewed with a graph. The graph shows the digital number on the y-axis and the different bands on the x-axis.

DATA (Current Workspace)   \03signatures\project1_bay_ss\bay_data_ss
RESULTS (Scratch Workspace)   \03signatures\project1_bay_ss\bay_results_ss

65

MAKING
SPATIAL
DECISIONS
USING GIS
AND REMOTE
SENSING

1

SPECTRAL
SIGNATURES

10. Add the ss_bay spreadsheet to the Table of Contents. (If you saved your data in a Microsoft Excel document, double-click ss_bay and add sheet1$.)

11. On the View menu, go to Graphs > Create Graph.

12. Graph the data in the ss_bay spreadsheet using the following parameters:
    a. For Graph type, select Vertical Line.
    b. For Layer/Table, select Sheet1$.
    c. Set the Y field to grass.
    d. Set the X label field to Band Name.
    e. Select the Show labels (marks) check box.
    f. Set the color to Custom green.
    g. Click Vertical line and change the name to grass.
    h. Go to Add > New Series.
    i. Repeat directions a–g for Concrete.
    j. Repeat directions a–g for Water.

13. Click Next and title the graph Selected Spectral Signatures.

14. Set Axis Properties Left to DN.

15. Set Axis Properties Bottom to Label as Bands, and click Finish.

16. Right-click the graph and export it as a .jpg file (JPEG) to your results folder. Call the graph table1_graph.

17. Save your map document as SS_Bay1. Do not close the map document.

Deliverable 1: A chart and graph of spectral signatures of concrete, grass, and water.

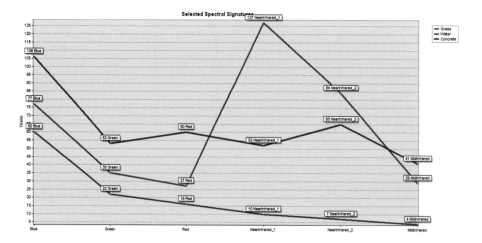

DATA (Current Workspace)   \03signatures\project1_bay_ss\bay_data_ss
RESULTS (Scratch Workspace)   \03signatures\project1_bay_ss\bay_results_ss

Q4 *Explain what DN represents.*

Q5 *Describe in detail and explain the spectral signatures of concrete, grass, and water.*

Q6 *Why does the spectral signature of grass spike in the NearInfrared_1 band?*

Q7 *Why is the spectral signature of water low?*

MAKING
SPATIAL
DECISIONS
USING GIS
AND REMOTE
SENSING

1

*SPECTRAL
SIGNATURES*

STEP 2: Compare spectral signatures of deciduous and coniferous forests

Vegetation has characteristics that distinguish it from other types of land cover. The reflectance of vegetation is low in the blue and red bands and peaks in the green band. The spectral response is much higher in the near-infrared region than in the visible region of the spectrum. This is because chlorophyll molecules in the leaves absorb energy in the blue and red bands but absorb little energy in the green band; this reflected green light gives many plants their characteristic color. Light is scattered and diffused by the walls of plant cells, causing plants to be strongly reflective in the near infrared. The spectral response of vegetation is also influenced by water content, time of year, and many other factors. Because this Landsat scene was taken in May 2006 in the southeastern United States, the deciduous trees should have their leaves out, and many agricultural crops will have emerged from the ground. You need to provide spectral signatures of deciduous trees, coniferous trees, and a crop area.

1. Close the Selected Spectral Signatures graph.

2. Turn off concrete, grass, and water.

3. Add dec_trees, con_trees, and crops from your data folder.

4. Using the same procedure as described in step 1, complete the following chart.

Band #	Band Name	Coniferous			Average	Deciduous			Average	Crops			Average
B10	Blue												
B20	Green												
B30	Red												
B40	NearInfrared_1												
B50	NearInfrared_2												
B70	MidInfrared												

DATA (Current Workspace)   \03signatures\project1_bay_ss\bay_data_ss
RESULTS (Scratch Workspace)   \03signatures\project1_bay_ss\bay_results_ss

67

5. Make a spreadsheet or a comma-delimited text file to store your data to import to ArcMap. The spreadsheet or comma-delimited text file should have the information shown in the chart.

MAKING
SPATIAL
DECISIONS
USING GIS
AND REMOTE
SENSING

1

SPECTRAL
SIGNATURES

Band #	Band Name	Coniferous	Deciduous	Crops
B10	Blue			
B20	Green			
B30	Red			
B40	NearInfrared_1			
B50	NearInfrared_2			
B70	MidInfrared			

6. Name the document veg_bay and save it to your results folder.

7. Repeat directions 11–16 in step 1 for veg_bay. The Y field should be Types of Vegetation.

8. Title the graph Vegetation of the Chesapeake Bay.

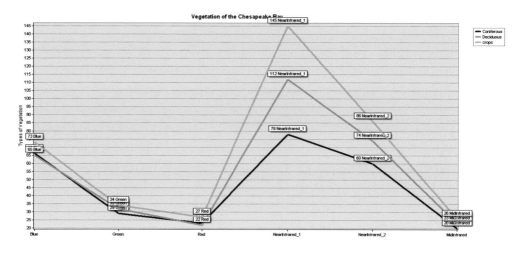

Q8 *Referring to the vegetation graph, which region of the spectrum (refer to the region by wavelength) shows the greatest reflectance for all vegetation?*

Q9 *Referring to the vegetation graph, which region of the spectrum (refer to the region by wavelength) shows the least reflectance for all vegetation?*

Q10 *Which vegetation type has the largest difference in range of spectral reflectance? Why?*

9. Right-click the graph and export it as a JPEG to your results folder. Call the graph table2_graph.

DATA (Current Workspace)   \03signatures\project1_bay_ss\bay_data_ss
RESULTS (Scratch Workspace)   \03signatures\project1_bay_ss\bay_results_ss

10. Save your map document.

Deliverable 2: A chart and graph comparing different types of vegetation.

STEP 3: Compare spectral signatures of clear water and turbid water

MAKING
SPATIAL
DECISIONS
USING GIS
AND REMOTE
SENSING

SPECTRAL
SIGNATURES

Certain water conditions can be assessed using spectral signatures. Clear water reflects very little light. However, turbid water and water that contains algae have an increase in reflectance. You will analyze two different types of water from the Chesapeake Bay by comparing their spectral signatures.

1. Close the Deciduous and Coniferous Tree graph.

2. Turn off the vegetation points.

3. Add Clear_water and turbid_water from your data folder.

4. Using the same procedure as described in step 1, complete the following chart.

Band #	Band Name	Clear Water			Average	Turbid Water			Average
B10	Blue								
B20	Green								
B30	Red								
B40	NearInfrared_1								
B50	NearInfrared_2								
B70	MidInfrared								

5. Make a spreadsheet or a comma-delimited text file to store your data to import to ArcMap. The spreadsheet or comma-delimited text file should have the following information. Save your spreadsheet as water_bay to your results folder.

Band #	Band Name	Clear Water	Turbid Water
B10	Blue		
B20	Green		
B30	Red		
B40	NearInfrared_1		
B50	NearInfrared_2		
B70	MidInfrared		

DATA (Current Workspace)   \03signatures\project1_bay_ss\bay_data_ss
RESULTS (Scratch Workspace)   \03signatures\project1_bay_ss\bay_results_ss

MAKING
SPATIAL
DECISIONS
USING GIS
AND REMOTE
SENSING

1

SPECTRAL
SIGNATURES

6. Repeat directions 11–16 in step 1. The Y field should be Types of Water.

7. Title the graph Clear and Turbid Water.

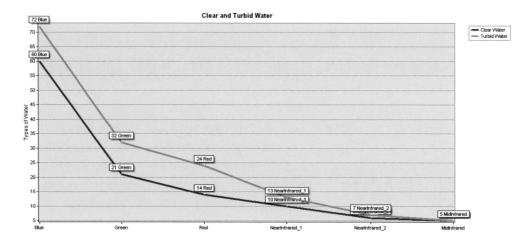

Q11 *Name two things that affect the spectral signature of water.*

Q12 *What might the suspended matter be in the turbid-water sample? Explain your answer.*

Q13 *What other qualities of water bodies would present difficulties when remotely sensed data is used?*

8. Right-click the graph and export it as a JPEG to your results folder. Call the graph table3_graph.

9. Save your map document.

Deliverable 3: A chart and graph comparing spectral signatures of clear water and turbid water.

MAKING
SPATIAL
DECISIONS
USING GIS
AND REMOTE
SENSING

PROJECT 2
Investigating spectral signatures of Las Vegas, Nevada

Scenario/problem

The Las Vegas urban planners have been told that spectral signatures have been developed for land categories. These land categories include native desert vegetation, disturbed desert land, paved surfaces, and vacant land. The planners do not have any understanding of spectral signatures and how they are used. They have asked PtD to provide material and training that would help them understand how they could use spectral signatures to make decisions about the Las Vegas urban area and the surrounding desert.

Objectives

For this project, you will do the following:
- Describe and understand the data collection needed to construct a spectral signature.
- Graph spectral signatures of different features.
- Analyze how events can affect spectral signatures by
  - comparing spectral signatures of different vegetation, and
  - comparing spectral signatures of clear and turbid water.

DATA (Current Workspace) \03signatures\project2_vegas_ss\vegas_data_ss
RESULTS (Scratch Workspace) \03signatures\project2_vegas_ss\vegas_results_ss

71

Deliverables

We recommend the following deliverables for this project:

1. A chart and graph of spectral signatures of concrete, grass, and water.
2. A chart and graph comparing spectral signatures of different types of vegetation.
3. A chart and graph comparing spectral signatures of clear and turbid water.

The questions for this project are both quantitative and qualitative. They identify key points that should be addressed in your analysis and presentation.

Keeping track of where your data and results are located is always a challenge. In these projects, we give the path to access the data (current workspace) and the path to store the results (scratch workspace) in a footnote on each page. The directions will specify whether the results go in the results folder or the results geodatabase.

Examine the data

This section was completed in module 1.

Organize and document your work

The data for this project is stored in the **\03signatures\project2_vegas_ss\vegas_data_ss** folder. Be sure to refer to project 1 and your process summary.

1. Set up the proper directory structure.

2. Create a process summary.

3. Document the map.

4. Set the environments as follows:
   a. Data frame coordinate system to Projected Coordinate Systems > UTM > WGS 1984 > Northern Hemisphere > WGS 1984 UTM Zone 11N.
   b. Current Workspace to \03signatures\project2_vegas_ss\vegas_data_ss.
   c. Scratch Workspace to \03signatures\project2_vegas_ss\vegas_results_ss.
   d. Output Coordinate System to Same as Display.

Analysis

Use the experience you acquired in project 1 to help you analyze spectral signatures for Las Vegas, Nevada.

MAKING
SPATIAL
DECISIONS
USING GIS
AND REMOTE
SENSING

2

SPECTRAL
SIGNATURES

STEP 1: Create three standard spectral signatures

1. Add the points concrete, grass, and water from your data folder.

2. Add L5039035_03520060512_MTL from your data folder.

MAKING
SPATIAL
DECISIONS
USING GIS
AND REMOTE
SENSING

2

SPECTRAL
SIGNATURES

Q1 *Where are the concrete points located?*

Q2 *Where is the grass located?*

Q3 *Where are the water points located?*

Q4 *Record the DN of each of the points for each of the bands in the chart on your worksheet. Average the three readings.*

Band #	Band Name	Grass			Average	Water			Average	Concrete			Average
B10	Blue												
B20	Green												
B30	Red												
B40	NearInfrared_1												
B50	NearInfrared_2												
B70	MidInfrared												

Q5 *Create a spreadsheet and graph of the average values.*

Band#	Band Name	grass	water	concrete
band1	Blue			
band2	Green			
band3	Red			
band4	NearInfrared_1			
band5	NearInfrared_2			
band7	MidInfrared			

Q6 *Explain the reflectance values (DN values).*

Q7 *Discuss in detail and explain the spectral signatures of concrete, grass, and water.*

Q8 *Why does the spectral signature of grass spike in band 4?*

Q9 *Why is the spectral signature of water low?*

Deliverable 1: A chart and graph of spectral signatures of concrete, grass, and water.

DATA (Current Workspace)   \03signatures\project2_vegas_ss\vegas_data_ss
RESULTS (Scratch Workspace)   \03signatures\project2_vegas_ss\vegas_results_ss

73

MAKING
SPATIAL
DECISIONS
USING GIS
AND REMOTE
SENSING

2

SPECTRAL
SIGNATURES

STEP 2: Compare spectral signatures of evergreen forest and shrub/scrub

1. Add the points evergreen and shrub/scrub from your data folder.

Q10 *Record the DN of each of the points for each of the bands in the chart on your worksheet. Average the three readings.*

Band #	Band Name	Evergreen			Average	Shrub_Scrub			Average
B10	Blue								
B20	Green								
B30	Red								
B40	NearInfrared_1								
B50	NearInfrared_2								
B70	MidInfrared								

Q11 *Create a spreadsheet and graph of the average values.*

Band#	Band Name	evergreen	shrub_scrub
band1	Blue		
band2	Green		
band3	Red		
band4	NearInfrared_1		
band5	NearInfrared_2		
band7	MidInfrared		

Q12 *Referring to the vegetation graph, which region of the spectrum (stated in bands) shows the greatest reflectance for evergreens?*

Q13 *Referring to the vegetation graph, which region of the spectrum (stated in bands) shows the greatest reflectance for shrub/scrub?*

Q14 *What is the difference in reflectance between the evergreens and the shrub/scrub?*

Deliverable 2: A chart and graph comparing evergreen trees and shrub/scrub.

STEP 3: Compare spectral signatures of clear water and turbid water

1. Add the points turbid_water and clear_water from your data folder.

MAKING
SPATIAL
DECISIONS
USING GIS
AND REMOTE
SENSING

2

SPECTRAL
SIGNATURES

Band #	Band Name	Clear Water			Average	Turbid Water			Average
B10	Blue								
B20	Green								
B30	Red								
B40	NearInfrared_1								
B50	NearInfrared_2								
B70	MidInfrared								

Band#	Band Name	clear_water	turbid_water
band1	Blue		
band2	Green		
band3	Red		
band4	NearInfrared_1		
band5	NearInfrared_2		
band7	MidInfrared		

Q15 *Name two things that affect the spectral signature of water.*

Q16 *What might the suspended matter be in the turbid-water sample? Explain your answer.*

Q17 *What other qualities of water bodies would present difficulties when remotely sensed data is used?*

Deliverable 3: A chart and graph comparing turbid and clear water.

DATA (Current Workspace)   \03signatures\project2_vegas_ss\vegas_data_ss
RESULTS (Scratch Workspace)   \03signatures\project2_vegas_ss\vegas_results_ss

75

MAKING
SPATIAL
DECISIONS
USING GIS
AND REMOTE
SENSING

3

SPECTRAL
SIGNATURES

PROJECT 3
On your own

Scenario/problem

You have worked through a project on spectral signatures for the Chesapeake Bay and repeated the analysis for Las Vegas. In this project, you will reinforce your skills by researching and analyzing a similar scenario entirely on your own. First, you must identify your study area and acquire the data for your analysis. You may want to study a local area. Refer to appendix A for directions on how to download your satellite imagery.

Research

Research the problem and answer the following questions:
1. What is the area of study?
2. What is the problem you are going to study?
3. What data is available?

Obtain the data

Do you have access to baseline data? Data and Maps for ArcGIS at http://www.esri.com/data/data-maps provides many of the layers of data that are needed for project work. Be sure to pay particular attention to the source of data and get the latest version. You can obtain data from the following sources:

- USGS Globalization Viewer at http://glovis.usgs.gov: Access to multiple sets of EROS satellite and aerial imagery.

- Census 2000 TIGER/Line Data at http://www.esri.com/tiger: Access to Census 2000 line data.
- Geospatial One-Stop at http://geo.data.gov: Web-based geospatial resources.
- The National Atlas at http://www.nationalatlas.gov: A range of products and geographic information about the United States.
- The National Map at http://nationalmap.gov/viewer.html: Data includes elevation, land cover, and topographic maps.

MAKING
SPATIAL
DECISIONS
USING GIS
AND REMOTE
SENSING

3

SPECTRAL
SIGNATURES

Workflow

After researching the problem and obtaining the data, you should do the following:

1. Write a brief scenario.

2. State the problem.

3. Define the deliverables.

4. Examine the metadata.

5. Set the directory structure, start your process summary, and document the map.

6. Decide what you need for the data frame coordinate system and the environments.
   a. What is the best projection for your work?
   b. Do you need to set a cell size or mask?

7. Start your analysis.

8. Prepare your presentation and deliverables.

9. Always remember to document your work in a process summary.

MODULE 4
LAND COVER

Introduction

For many remotely sensed images, especially from Landsat, a primary purpose is to determine what is present on the ground in the scene. This is called land-cover analysis, and it focuses on identifying areas where water, bare ground, forest, developed land, or some other land type is present. There are published datasets that provide land cover, but you should always verify such data for your particular scene. In this module, you will learn some basic techniques for verification, and then assess the accuracy of the published datasets.

Scenarios in this module

- Comparing digitized land cover to USGS land-cover classes in Loudoun County, Virginia
- Comparing digitized land cover to USGS land-cover classes near Lake Mohave
- On your own

Student worksheets

The student worksheet files can be found on the Maps and Data DVD.

Project 1: Loudoun County student sheet

- File name: 04a_landcover_worksheet
- Location: \Student_Worksheets\04landcover

Project 2: Lake Mohave student sheet

- File name: 04b_landcover_worksheet
- Location: \Student_Worksheets\04landcover

MAKING
SPATIAL
DECISIONS
USING GIS
AND REMOTE
SENSING

1

PROJECT 1
Comparing digitized land cover to USGS land-cover classes in Loudoun County, Virginia

Background

There are two common methods for deriving land-cover information. The first method involves digitizing land-cover features from aerial photography or high spatial resolution satellite imagery, and the second method involves the classification of remotely sensed data. Comprehensive land-cover studies involve a combination of digitizing and spectral classification, which allows the analyst to use both vector and raster data.

Vector data is obtained from any appropriate image in which the resolution permits the features of interest to be discerned clearly. The data is based on lines and areas derived by drawing (digitizing) boundaries around features (see the accompanying figure) and defining or labeling the categories with attributes. In this module, online orthophotographs will be used as a basemap for digitizing. Orthophotographs are aerial photographs in a digital format that have had a correction made so that the image is correctly georeferenced.

DATA (Current Workspace)   \04landcover\project1_loudoun\loudoun_data
RESULTS (Scratch Workspace)   \04landcover\project1_loudoun\loudoun_results

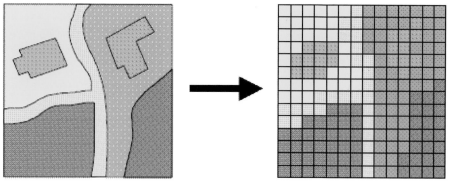

Comparing vector and raster data.

MAKING
SPATIAL
DECISIONS
USING GIS
AND REMOTE
SENSING

1

LAND COVER

A raster dataset is composed of pixels with information about the earth's surface. In this case, each pixel represents a designation or attribute of land cover. In this module, we will be using the National Land Cover Database of 2006. This dataset was produced by assigning each pixel in a Landsat TM or Enhanced Thematic Mapper Plus (ETM+) image to one of 16 land-cover classes using a procedure known as unsupervised classification. This procedure is discussed in module 5. Data at a spatial resolution of 30 meters is available for all US states, except Hawaii and Alaska. This dataset was produced by the Multi-Resolution Land Characteristics Consortium (MRLC), which includes such agencies as the Environmental Protection Agency (EPA), the National Oceanic and Atmospheric Administration (NOAA), the US Geological Survey (USGS), the US Forest Service (USFS), and NASA. The 16-class land-cover classification legend for the National Land Cover Database (NLCD) at http://www.mrlc.gov/nlcd06_leg.php is shown in the figure.

NLCD Land Cover Classification Legend

- 11 Open Water
- 12 Perennial Ice/ Snow
- 21 Developed, Open Space
- 22 Developed, Low Intensity
- 23 Developed, Medium Intensity
- 24 Developed, High Intensity
- 31 Barren Land (Rock/Sand/Clay)
- 41 Deciduous Forest
- 42 Evergreen Forest
- 43 Mixed Forest
- 51 Dwarf Scrub*
- 52 Shrub/Scrub
- 71 Grassland/Herbaceous
- 72 Sedge/Herbaceous*
- 73 Lichens*
- 74 Moss*
- 81 Pasture/Hay
- 82 Cultivated Crops
- 90 Woody Wetlands
- 95 Emergent Herbaceous Wetlands

* Alaska only

Land-cover categories from the National Land Cover Database.

DATA (Current Workspace)   \04landcover\project1_loudoun\loudoun_data
RESULTS (Scratch Workspace) \04landcover\project1_loudoun\loudoun_results

81

MAKING
SPATIAL
DECISIONS
USING GIS
AND REMOTE
SENSING

LAND COVER

Scenario/problem

Land-cover data is commonly used to assess the environmental quality of an area, such as the degree to which features such as forests are fragmented by development. However, decisions have to be made about the land-cover data of the study area. The Chesapeake Bay Foundation has also heard the adage that "raster is faster, but vector is better." The foundation needs a quantitative analysis before it can determine when to use vector data and when to use raster data to study land cover in the Chesapeake Bay watershed. Your job is to compare vector and raster data and present CBF with quantitative results.

Objectives

Land-cover accuracy assessment of thematic maps derived from image classification involves the calculation and comparison of areas of different types of land cover in these maps. You have been asked to produce the following elements to be included in the assessment of land-cover accuracy for Loudoun County, Virginia:

- A digital version of land cover using aerial photography
- A calculation of land areas using standard land classification codes
- A comparison of calculated land cover with standard 2006 land cover derived from multi-spectral imagery

Deliverables

We recommend the following deliverables for this project:

1. A basemap of Beaverdam Creek Reservoir.
2. A sample map of Beaverdam Creek Reservoir with developed and open spaces, water and wetlands, and vegetation digitized from aerial photography. The map must also include classification codes and land area.
3. A map of 2006 land cover displayed with the categories of water, developed, and vegetation.
4. A chart comparing the accuracy of the two methods of land-cover classification.

The questions for this project are both quantitative and qualitative. They identify key points that should be addressed in your analysis and presentation.

Keeping track of where your data and results are located is always a challenge. In these projects, we give the path to access the data (current workspace) and the path to store the results (scratch workspace) in a footnote on each page. The directions will specify whether the results go in the results folder or the results geodatabase.

Examine the data

This section was completed in module 1.

Organize and document your work

MAKING
SPATIAL
DECISIONS
USING GIS
AND REMOTE
SENSING

1

LAND COVER

The following preliminary steps are essential to a successful geospatial analysis.

Examine the directory structure

In a geospatial project, you must carefully keep track of the data and your calculations. You will work with a number of different files, and it is important to keep them organized so you can easily find them. The best way to do this is to have a folder for your project that contains a data folder. For this project, the folder named **\04landcover\project1_loudoun\loudoun_data** will be your project folder. Make sure it is stored in a place where you have write access.

You can store your data inside the results folder. The results folder already contains an empty geodatabase named **loudoun_results** for this purpose. Store your map documents in the **loudoun_results** folder.

Folder structure:
    04landcover
          project1_loudoun
              loudoun_data
                  bay.gdb
              loudoun_results
                  loudoun_results.gdb

Create a process summary

The process summary is simply a list of the steps you used to do your analysis. We suggest using a simple text document for your process summary. Keep adding to it as you do your work to avoid forgetting any steps. The following list shows an example of the first few entries in a process summary:

1. Produce a map of the Loudoun County area.
2. Digitize the designated land-cover areas.

A process summary is essentially a work log. It will allow you to remember the steps in your data analysis at a later time or communicate them with others who might need to reproduce your work.

DATA (Current Workspace)   \04landcover\project1_loudoun\loudoun_data
RESULTS (Scratch Workspace)   \04landcover\project1_loudoun\loudoun_results

83

Document the map

1. Start ArcMap and add descriptive properties to your map document properties.

2. Be sure to select the pathnames check box to store relative pathnames to all your data.

MAKING
SPATIAL
DECISIONS
USING GIS
AND REMOTE
SENSING

LAND COVER

Set the environments

1. On the View menu, click Data Frame Properties. Set the map projection to Projected Coordinate Systems > UTM > WGS 1984 > Northern Hemisphere > WGS 1984 UTM Zone 18N.

2. Set the Current Workspace to \04landcover\project1_loudoun\loudoun_data.

3. Set the Scratch Workspace to \04landcover\project1_loudoun\loudoun_results.

4. For Output Coordinate System, select Same as Display.

5. For Processing Extent, navigate to the bayfeatures dataset and set the extent to BD_study_area. For the Snap Raster, navigate to the data folder and select LC_rec.

6. Click OK and save your map document as LC_Loudoun1.

Analysis

Once you have examined the data, completed the map documentation, and set the environments, you are ready to begin the analysis and complete the data displays you need to address the problem. For this project, you have been asked to do a land-cover assessment for a specific section of Loudoun County, Virginia. To put the problem in context, you will start by creating a basemap of Beaverdam Creek Reservoir, the assigned study area in Loudoun County.

STEP 1: Create a basemap of Beaverdam Creek Reservoir in Loudoun County

1. Add Beaverdam from your data folder.

2. Add the basemap Imagery with Labels using the Add Data button on the Standard toolbar.

3. Switch to Layout View and create a map document with appropriate scale, north arrow, and title.

4. Save your map document.

DATA (Current Workspace)   \04landcover\project1_loudoun\loudoun_data
RESULTS (Scratch Workspace) \04landcover\project1_loudoun\loudoun_results

5. Save the map document a second time as LC_Loudoun2.

Q1 ***Write a description of land cover around Beaverdam Creek Reservoir.***

Deliverable 1: A basemap of Beaverdam Creek Reservoir.

MAKING
SPATIAL
DECISIONS
USING GIS
AND REMOTE
SENSING

STEP 2: Digitize land-cover areas around Beaverdam Creek Reservoir

1. Open LC_Loudoun2 and switch to Data View.

2. Remove Beaverdam.

3. Add BD_study_area from your data folder. This study area represents the small area where vector and raster data will be compared. This area was chosen to reduce digitizing time.

4. Make the study area hollow.

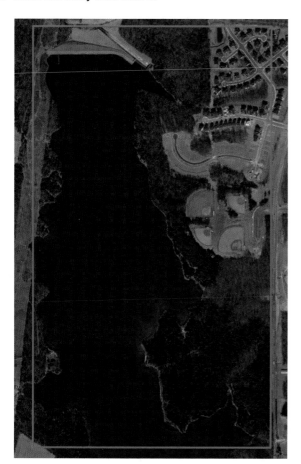

DATA (Current Workspace)   \04landcover\project1_loudoun\loudoun_data
RESULTS (Scratch Workspace)   \04landcover\project1_loudoun\loudoun_results

85

You are going to digitize the study area into three types of land cover:

- Developed and open spaces
- Water and wetlands
- Vegetation

MAKING
SPATIAL
DECISIONS
USING GIS
AND REMOTE
SENSING

LAND COVER

The easiest land feature to identify is water, the second easiest is the forest, and the hardest is the combination of open space and developed land cover.

5. Add the feature class named digitize from your data folder. This is an empty feature class that you will populate by digitizing the three types of land cover.

6. Turn on the Editor toolbar.

7. Go to Editor > Start Editing.

8. From the Editor drop-down menu, turn on the Snapping toolbar. Go to Snapping > Snap To Sketch. Be sure that the Editor toolbar is located where it is easily accessible.

9. Open the Create Features window by clicking Create Features on the Editor toolbar. This is the last icon on the right side of the toolbar. This sets up the editing environment so that you can begin to digitize your three land-cover features. The Create Features window appears on the right side of your screen.

10. You are now ready to start digitizing the first land-cover feature. Select the feature class digitize in the Create Features window. At the bottom of the Create Features window, you will see Construction Tools. Select the Polygon construction tool.

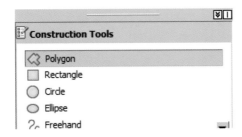

DATA (Current Workspace)   \04landcover\project1_loudoun\loudoun_data
RESULTS (Scratch Workspace)   \04landcover\project1_loudoun\loudoun_results

11. Before you start digitizing the lake, go to Editor > Options. Select the Show feature construction toolbar check box. The feature construction mini-toolbar provides quick access to some of the most commonly used editing commands. Three tools are used in this project:
   - Straight Segment: The default method to digitize the vertices of a line or polygon feature.
   - Trace Segment: A quick and accurate way of creating new segments that follow the shapes of other features. You can trace directly on top of a feature.
   - Undo Add Vertex: This tool allows you to delete vertices.

12. Start editing and digitize the lake within the study area.

13. Double-click to stop digitizing.

MAKING
SPATIAL
DECISIONS
USING GIS
AND REMOTE
SENSING

1

LAND COVER

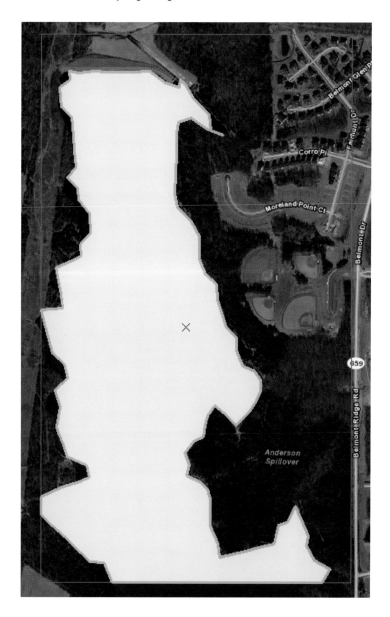

DATA (Current Workspace)   \04landcover\project1_loudoun\loudoun_data
RESULTS (Scratch Workspace)   \04landcover\project1_loudoun\loudoun_results

87

MAKING
SPATIAL
DECISIONS
USING GIS
AND REMOTE
SENSING

1

LAND COVER

14. Digitize the vegetation. When you are near the water polygon you have already digitized, you can turn on the Trace tool on the feature construction toolbar. This allows you to trace along the edge of the already digitized water polygon.

15. Digitize the developed and open spaces features in the raster.

16. Open the digitize attribute table. There are no values in the Type field. If you click Null, a drop-down menu will appear and you can select which land-cover type is represented.

	OBJECTID *	SHAPE *	Type	SHAPE_Length	SHAPE_Area
▶	18	Polygon	Water	5598.350365	671160.836703
	21	Polygon	Vegetation	5606.469852	424504.293909
	24	Polygon	Vegetation	732.811767	23217.334323
	25	Polygon	Vegetation	1291.296503	32566.092453
	31	Polygon	Vegetation	1311.009952	30677.440575
	32	Polygon	Developed	2554.999773	82097.769547
	33	Polygon	Developed	2535.564171	257656.462065

digitize

17. On the Editor toolbar, go to Stop Editing > Save Your Edits.

18. Symbolize digitize by unique values with the value field Type. Be sure to click Add All Values. This will display each land-cover type individually.

Your study area should now look similar to the following screen capture.

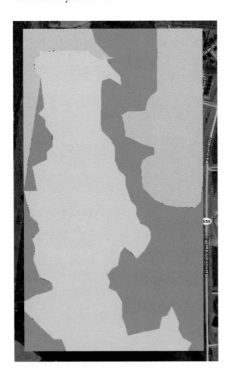

DATA (Current Workspace)   \04landcover\project1_loudoun\loudoun_data
RESULTS (Scratch Workspace)   \04landcover\project1_loudoun\loudoun_results

19. The last step for the vector feature data analysis is to record the area of each type of land cover. There is only one polygon with water as the land cover. Record the SHAPE_Area of water below. Because the map projection is UTM, the area is given in square meters.

water =

Select all the polygons with developed land cover and right-click SHAPE_Area. Click Statistics and record the sum of the areas of these polygons.

developed =

Repeat this direction to generate the area with vegetation as the land cover.

vegetation =

You can evaluate your digitizing in various ways. You can set the transparency to 65%, and then compare with the image beneath the digitized layer. You can turn on the Effects toolbar and swipe the digitized layer. Or you can simply turn the digitized layer on and off.

Q2 *Evaluate the accuracy of your digitizing using one of the methods described. Can you see areas that you missed or areas that you included in the wrong land-cover class?*

20. Switch to Layout View and create an appropriate map document.

21. Save your map document.

22. Save the map document again as LC_Loudoun3.

Deliverable 2: A sample map of Beaverdam Creek Reservoir with developed and open spaces, water and wetlands, and vegetation digitized from aerial photography. The map must also include classification codes and land area.

MAKING
SPATIAL
DECISIONS
USING GIS
AND REMOTE
SENSING

1

LAND COVER

DATA (Current Workspace)   \04landcover\project1_loudoun\loudoun_data
RESULTS (Scratch Workspace)   \04landcover\project1_loudoun\loudoun_results

89

STEP 3: Reclassify 2006 land cover

1. Open the map document LC_Loudoun3.

2. Switch to Data View.

MAKING
SPATIAL
DECISIONS
USING GIS
AND REMOTE
SENSING

3. To make navigating easier, turn off the basemap layer but leave it in the Table of Contents.

4. Add LC_rec from your data folder.

Q3   *How can you tell this is raster data?*

Q4   *Zoom in on an individual pixel and use the Measuring tool to measure the spatial dimensions of a pixel. Record your findings in meters.*

Q5   *Calculate the area of a single pixel.*

5. Right-click LC_rec and go to Properties > Symbology. If you are asked to create the attribute table, click Yes. Display by unique values.

6. Select appropriate colors for each land cover and label the land-cover classes.

Q6   *Can you easily identify the land-cover classes?*

In the attribute table of LC_rec, there is a field named Count. This field gives the number of pixels in each land-cover class. Knowing the number of pixels in each land-cover class and the area of a single pixel lets you calculate the area of each type of land cover.

7. With the attribute table of LC_rec open, click the Table Options button in the upper-left corner and click Add Field.
   a. Name the field Area.
   b. For Type, select Float.
   c. Click OK.

8. Right-click the Area field heading and click Field Calculator. Enter the following formula to calculate the area:
   [Count] *30 *30.

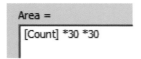

DATA (Current Workspace)   \04landcover\project1_loudoun\loudoun_data
RESULTS (Scratch Workspace)   \04landcover\project1_loudoun\loudoun_results

9. Record the area in square meters of each land-cover type:

water =

developed =

vegetation =

10. Switch to Layout View and create an appropriate map document. The map document should show both the vector and the raster land cover. You may show both layers, symbolized differently, in the same data frame, or you may choose to make a map document with two data frames.

11. Save your map document.

Deliverable 3: A map of 2006 land cover displayed with the categories of water, developed, and vegetation.

STEP 4: Compare the accuracy of the two methods

Historically, raster calculations involving land cover are faster to compute than vector calculations. However, vector measurements of land cover are usually more accurate, typically because the vector data has better spatial resolution (remember the raster data here has a resolution of 30 meters).

Q7   *Enter the values obtained for the vector and raster methods of area calculation for the different land-cover types in the chart on your worksheet. Calculate the difference between the two values.*

Land-cover type	Orthophotograph-based (vector) land cover (m²)	NLCD-based land cover (m²)	Difference (m²)
water			
developed			
vegetation			
Total			

MAKING
SPATIAL
DECISIONS
USING GIS
AND REMOTE
SENSING

1

LAND COVER

DATA (Current Workspace)   \04landcover\project1_loudoun\loudoun_data
RESULTS (Scratch Workspace)   \04landcover\project1_loudoun\loudoun_results

91

1. Turn on the Effects toolbar and use the Swipe tool to compare the orthophotograph (raster) dataset to the NLCD (vector) dataset.

Q8 *Where do you see discrepancies between the vector and raster datasets?*

Q9 *What could be a reason for this difference in area?*

MAKING
SPATIAL
DECISIONS
USING GIS
AND REMOTE
SENSING

1

LAND COVER

2. Zoom to the edge of the raster/vector study area. When you zoom to the edge of the study area, you see that the vector file and the raster file do not align with each other. The vector file has a defined edge, and the raster file is pixelated due to the structural building blocks of a raster.

Q10 *Do the vector and raster datasets align? Why or why not?*

From the work you have done with rasters and vectors in this project, answer the following questions.

Q11 *Which data format would be best suited for defining boundaries?*

Q12 *Which data format would be best suited for representing the natural landscape?*

Q13 *Would pixel size influence the accuracy or validity of raster land cover?*

Q14 *If there is a file size limit, which data would be the most appropriate to use?*

Q15 *Which type of data (raster or vector) would you choose to represent the following:*
   a. Elevation
   b. Tax parcels
   c. Deciduous trees
   d. Streets

Deliverable 4: A chart comparing the accuracy of the two methods of land-cover classification.

DATA (Current Workspace)   \04landcover\project1_loudoun\loudoun_data
RESULTS (Scratch Workspace)   \04landcover\project1_loudoun\loudoun_results

MAKING
SPATIAL
DECISIONS
USING GIS
AND REMOTE
SENSING

PROJECT 2
Comparing digitized land cover to USGS land-cover classes near Lake Mohave

Scenario/problem

Cottonwood Cove Resort and Marina on Lake Mohave is located in Nevada between Laughlin, Nevada, and Bullhead City, Arizona. It has calm water and offers spectacular scenery. The resort is treated as an incorporated city. In order to ensure that the resort and marina has equitable real estate property taxes, the Assessor's Office has hired PtD to do a comprehensive study of land cover. It has been using digitized land-cover information. However, PtD would like to switch to the easier raster land-cover property assessment. You have been hired to do a comprehensive comparison between the two land-cover evaluations.

DATA (Current Workspace)   \04landcover\project2_mohave\mohave_data
RESULTS (Scratch Workspace)   \04landcover\project2_mohave\mohave_results

93

Cottonwood Cove Resort and Marina on Lake Mohave

MAKING
SPATIAL
DECISIONS
USING GIS
AND REMOTE
SENSING

LAND COVER

0   0.175   0.35   0.7 Miles

Objectives

Land-cover assessment using thematic maps derived from image classification involves the assessment of accuracy of the maps. You have been asked to produce the following to help assess the accuracy of land cover for Lake Mohave:

- A digital version of land cover using aerial photography
- A calculation of land areas using standard land classification codes
- A comparison of the land cover to standard 2006 land cover derived from multispectral imagery

Deliverables

We recommend the following deliverables for this exercise:

1. A basemap of Cottonwood Cove Resort and Marina in relation to Lake Mohave.

DATA (Current Workspace)   \04landcover\project2_mohave\mohave_data
RESULTS (Scratch Workspace)   \04landcover\project2_mohave\mohave_results

2. A sample map of Cottonwood Cove Resort and Marina with developed and open spaces, water and wetlands, and vegetation digitized from aerial photography. The map must also include classification codes and land area.

3. A map showing the reclassification of 2006 land cover into the categories of water/wetlands, forest, and pasture/crop.

4. A chart comparing the accuracy of the two methods of land-cover classification.

The questions for this project are both quantitative and qualitative. They identify key points that should be addressed in your analysis and presentation.

Keeping track of where your data and results are located is always a challenge. In these projects, we give the path to access the data (current workspace) and the path to store the results (scratch workspace) in a footnote on each page. The directions will specify whether the results go in the results folder or the results geodatabase.

Examine the data

This section was completed in module 1.

Organize and document your work

The data for this project is stored in the **\04landcover\project2_mohave\mohave_data** folder. Be sure to refer to project 1 and your process summary.

1. Set up the proper directory structure.

2. Create a process summary.

3. Document the map.

4. Set the environments:
   a. Data frame coordinate system to Projected Coordinate Systems > UTM > WGS 1984 > Northern Hemisphere > WGS 1984 UTM Zone 11N
   b. Current Workspace to \04landcover\project2_mohave\mohave_data
   c. Scratch Workspace to \04landcover\project2_mohave\mohave_results
   d. Output Coordinate System to Same as Display

MAKING
SPATIAL
DECISIONS
USING GIS
AND REMOTE
SENSING

2

LAND COVER

DATA (Current Workspace)   \04landcover\project2_mohave\mohave_data
RESULTS (Scratch Workspace)   \04landcover\project2_mohave\mohave_results

95

Analysis

STEP 1: Create a basemap of Cottonwood, Nevada, near Lake Mohave

MAKING
SPATIAL
DECISIONS
USING GIS
AND REMOTE
SENSING

The study_area layer in the data folder contains Cottonwood Cove Resort and Marina.

Q1   ***Write a description of land cover for Cottonwood Cove Resort and Marina.***

Deliverable 1: A basemap of Cottonwood Cove Resort and Marina in relation to Lake Mohave.

STEP 2: Digitize land-cover areas around Cottonwood Cove Resort and Marina

Q2   ***Using one of the methods described, evaluate your digitizing. Can you see areas that you missed or areas that you included in the wrong land-cover class?***

Deliverable 2: A sample map of Cottonwood Cove Resort and Marina with developed and open spaces, water and wetlands, and vegetation digitized from the imagery basemap. The map must also include classification codes and land areas.

Q3   ***How can you tell this is raster data?***

Q4   ***Zoom in on a single pixel and with the Measuring tool measure the spatial dimensions of a pixel. Record your findings in meters.***

Q5   ***Calculate the area of one pixel.***

Q6   ***Can you identify the land-cover classes?***

STEP 3: Reclassify 2006 land cover

Deliverable 3: A map showing the reclassification of 2006 land cover into the categories of water/wetlands, forest, and pasture/crop.

DATA (Current Workspace)   \04landcover\project2_mohave\mohave_data
RESULTS (Scratch Workspace)   \04landcover\project2_mohave\mohave_results

STEP 4: Compare the accuracy of the two methods

Q7 *Enter the values obtained for the vector and raster methods of area calculation in the chart on your worksheet. Calculate the difference between the two values.*

MAKING
SPATIAL
DECISIONS
USING GIS
AND REMOTE
SENSING

2

LAND COVER

Land-cover type	Vector area (m²)	Raster area (m²)	Difference (m²)
water			
developed			
vegetation			
Total			

Q8 *What could be a reason for the difference in area?*

Hint: Zoom to the edge of the raster/vector study areas.

Deliverable 4: A chart comparing the accuracy of the two methods of land-cover classification.

MAKING
SPATIAL
DECISIONS
USING GIS
AND REMOTE
SENSING

3

PROJECT 3
On your own

Scenario/problem

You have worked through a project on land cover for the Chesapeake Bay and repeated the analysis for Lake Mohave. In this project, you will reinforce your skills by researching and analyzing a similar scenario entirely on your own. First, you must identify your study area and acquire the data for your analysis. You may want to study a local area. Refer to appendix A for directions on how to download your satellite imagery.

Research

Research the problem and answer the following questions:

1. What is the area of study?
2. What is the problem you are going to study?
3. What data is available?

Obtain the data

Do you have access to baseline data? Data and Maps for ArcGIS at http://www.esri.com/data/data-maps provides many of the layers of data that are needed for project work. Be sure to pay particular attention to the source of data and get the latest version. You can obtain data from the following sources:

- USGS Globalization Viewer at http://glovis.usgs.gov: Access to multiple sets of EROS satellite and aerial imagery.

- Census 2000 TIGER/Line Data at http://www.esri.com/tiger: Access to Census 2000 line data.
- Geospatial One-Stop at http://geo.data.gov: Web-based geospatial resources.
- The National Atlas at http://www.nationalatlas.gov: A range of products and geographic information about the United States.
- The National Map at http://nationalmap.gov/viewer.html: Data includes elevation, land cover, and topographic maps.

MAKING
SPATIAL
DECISIONS
USING GIS
AND REMOTE
SENSING

Workflow

After researching the problem and obtaining the data, you should do the following:

1. Write a brief scenario.

2. State the problem.

3. Define the deliverables.

4. Examine the metadata.

5. Set the directory structure, start your process summary, and document the map.

6. Decide what you need for the data frame coordinate system and the environments.
   a. What is the best projection for your work?
   b. Do you need to set a cell size or mask?

7. Start your analysis.

8. Prepare your presentation and deliverables.

9. Always remember to document your work in a process summary.

MODULE 5
UNSUPERVISED CLASSIFICATION

Introduction

The large size of remote sensing images makes it important to be able to automate the process of determining the land cover represented by different pixels in the scene. In this module and the next, you will learn two different methods for having ArcMap perform this function: unsupervised classification and supervised classification. In this module, you will focus on the unsupervised classification and comparison of different watersheds.

Scenarios in this module

- Calculating unsupervised classification of the Chesapeake Bay
- Calculating unsupervised classification of Las Vegas, Nevada
- On your own

Student worksheets

The student worksheet files can be found on the Maps and Data DVD.

Project 1: Chesapeake Bay student sheet
- File name: 05a_unsupervised_worksheet
- Location: \Student_Worksheets\05unsupervised

Project 2: Las Vegas student sheet
- File name: 05b_unsupervised_worksheet
- Location: \Student_Worksheets\05unsupervised

MAKING
SPATIAL
DECISIONS
USING GIS
AND REMOTE
SENSING

1

UNSUPERVISED
CLASSIFICATION

PROJECT 1

Calculating unsupervised classification of the Chesapeake Bay

Background

A primary use of Landsat imagery is to interpret and classify features. Analysis schemes that synthesize the data collected by the sensors at different wavelengths (spectral bands) aid in this process. One type of analysis involves categorizing the imagery into different land-cover types. There are different ways to do this categorization. In this project, you will focus on unsupervised classification, a method where the software does the work of grouping different pixels into similar types of land cover. The analyst has only to specify the number of classes and the number of bands to use. The software generates the classes based on the likeness of the DN values of the pixels in the scene.

Scenario/problem

The Chesapeake Bay Foundation is doing a study of the variations in land cover of different watersheds in the metropolitan area around and between Washington, DC, and Baltimore, Maryland. It has identified three watersheds for analysis: Patuxent, Severn, and Middle Potomac-Anacostia-Occoquan. CBF has asked your firm to classify 2006 Landsat TM imagery into categories of land cover. It wants to use this land-cover classification in a research proposal to request federal money for bay watershed restoration (cf. Goetz et al. 2004; Yang et al. 2003).

DATA (Current Workspace)   \05unsupervised\project1_bay_unsup\bay_data_unsup
RESULTS (Scratch Workspace)   \05unsupervised\project1_bay_unsup\bay_results_unsup

It would like to analyze the watersheds based on the following classes: water, developed, agricultural, and forest (note that these categories come from the Anderson land-use and land-cover classification system, a standard used in the United States, cf. Anderson et al. 1976). It would like graphs showing the percentages of each of these land types to analyze the type of protection and restoration needed for each watershed.

MAKING
SPATIAL
DECISIONS
USING GIS
AND REMOTE
SENSING

1

UNSUPERVISED
CLASSIFICATION

Objectives

For this project, you will use Landsat imagery to do the following:

- Perform unsupervised classification with postprocessing cleanup of Landsat imagery.
- Combine the classes of the unsupervised classification into the categories of water, developed, forest, crop/pasture, and wetlands.
- Compare land cover of different watersheds.
- Use the classification to make recommendations for watershed protection and restoration.

Deliverables

We recommend the following deliverables for this project:

1. A map showing the Severn, Patuxent, and Middle Potomac-Anacostia-Occoquan watersheds with streams, places, and highways identified.
2. An unsupervised classification of the complete watershed area. The classes of land should include water, developed, forest, crop/pasture, and wetlands.
3. A comparison graph with percentages of land cover for each of the three watersheds.
4. A written analysis using maps and graphs describing the type of protection and restoration needed for each watershed.

The questions for this project are both quantitative and qualitative. They identify key points that should be addressed in your analysis and presentation.

Keeping track of where your data and results are located is always a challenge. In these projects, we give the path to access the data (current workspace) and the path to store the results (scratch workspace) in a footnote on each page. The directions will specify whether the results go in the results folder or the results geodatabase.

Examine the data

This section was completed in module 1.

Organize and document your work

The following preliminary steps are essential to a successful geospatial analysis.

DATA (Current Workspace)   \05unsupervised\project1_bay_unsup\bay_data_unsup
RESULTS (Scratch Workspace)   \05unsupervised\project1_bay_unsup\bay_results_unsup

103

Examine the directory structure

In a geospatial project, you must carefully keep track of the data and your calculations. You will work with a number of different files, and it is important to keep them organized so you can easily find them. The best way to do this is to have a folder for your project that contains a data folder. For this project, the folder named **\05unsupervised\project1_bay_unsup\bay_data_unsup** will be your project folder. Make sure it is stored in a place where you have write access.

MAKING
SPATIAL
DECISIONS
USING GIS
AND REMOTE
SENSING

1

UNSUPERVISED
CLASSIFICATION

You can store your data inside the results folder. The results folder already contains an empty geodatabase named **unsupbay_results** for this purpose. Save your map documents to the **bay_results_unsup** folder.

Folder structure:
    05unsupervised
        project1_bay_unsup
            bay_data_unsup
                bay.gdb
            bay_results_unsup
                unsupbay_results.gdb

Create a process summary

The process summary is simply a list of the steps you used to do your analysis. We suggest using a simple text document for your process summary. Keep adding to it as you do your work to avoid forgetting any steps. The following list shows an example of the first few entries in a process summary:

1. Use the Select by Rectangle tool to select the three watersheds.
2. Add the metadata for the Landsat scene.
3. Extract by mask the Landsat scene to the watersheds.

Document the map

1. Start ArcMap and add descriptive properties to your map document properties.

2. Be sure to select the pathnames check box to store relative pathnames to all your data.

Set the environments

1. On the View menu, click Data Frame Properties. Set the map projection to Projected Coordinate Systems > UTM > WGS 1984 > Northern Hemisphere > WGS 1984 UTM Zone 18N.

DATA (Current Workspace)   \05unsupervised\project1_bay_unsup\bay_data_unsup
RESULTS (Scratch Workspace)   \05unsupervised\project1_bay_unsup\bay_results_unsup

2. Set the Current Workspace to \05unsupervised\project1_bay_unsup\bay_data_unsup.

3. Set the Scratch Workspace to \05unsupervised\project1_bay_unsup\bay_results_unsup.

4. For Output Coordinate System, select Same as Display.

5. Set the Processing Extent to the same as sel_sheds.

6. Click OK and save your project as UnsupBay1.

MAKING
SPATIAL
DECISIONS
USING GIS
AND REMOTE
SENSING

1

UNSUPERVISED
CLASSIFICATION

Analysis

Once you have examined the data, completed the map documentation, and set the environments, you are ready to begin the analysis and complete the displays you need to address the problem. For this project, you have been asked to classify Landsat imagery for an area that encompasses three watersheds. The creation of a basemap displaying the watersheds is a good place to begin your analysis.

STEP 1: Identify and map designated watersheds

1. Add L5015033_03320060504_MTL from the data folder.

2. Go to Properties > Symbology and click Yes when asked if you want to make a histogram. Refresh the band combination red, green, and blue by selecting the channels again.

3. Add sel_sheds, shed_rivers, and shed_highways from your data folder.

Now that the feature classes have been created, the Landsat scene needs to be clipped to the watershed boundaries.

4. Select all the polygons in the sel_sheds layer.

5. On the Windows menu, click Image Analysis to open the window.

6. Highlight the MTL file, and then click the Clip icon. The Clip icon is the first icon in the Processing panel of the Image Analysis window.

7. This creates a temporary file called Clip_Multispectral_L5015033_03320060504_MTL.

8. Remove Multispectral_L5015033_03320060504_MTL.

DATA (Current Workspace)   \05unsupervised\project1_bay_unsup\bay_data_unsup
RESULTS (Scratch Workspace)   \05unsupervised\project1_bay_unsup\bay_results_unsup

105

You will be working with a lot of temporary files to do the unsupervised classification. You will not be required to save any of these files until the final product. However, because they are temporary files, you need to finish the exercise in one session on your computer or else the temporary files will be lost when you quit ArcMap.

MAKING
SPATIAL
DECISIONS
USING GIS
AND REMOTE
SENSING

1

UNSUPERVISED
CLASSIFICATION

9. Clear the selection of the sel_sheds layer.

10. Right-click shed_highways and go to Properties > Symbology and import the layer file highways from your data folder.

11. Make shed_rivers an appropriate color.

12. Make sel_sheds hollow and label the watersheds. Select a color for the sel_sheds outline that makes the watersheds look distinct.

13. Save your map document.

14. Save the map document a second time as unsupBay2.

Q1   ***Using different band combinations, write a short analysis of the land-cover patterns in each watershed.***

Q2   ***How do you think the land cover affects the health of each watershed?***

Deliverable 1: A map showing the Severn, Patuxent, and Middle Potomac-Anacostia-Occoquan watersheds with streams, places, and highways identified.

STEP 2: Classify Iso cluster unsupervised

The goal of pixel-based classification is to assign each pixel in a study area to a single class or category. Land-use type is a very common example of a class or category. In an unsupervised classification, the assignment of the class or cluster of each location is dependent on the statistics calculated by ArcMap. Unsupervised classification does not require any information from the analyst. A class or cluster corresponds to a meaningful grouping of pixels based on their DN values. The classes of land in this unsupervised classification should include water, developed, forest, crop/pasture, and wetlands.

1. Open the UnsupBay2 map document and go to Data View.

2. Remove all layers except Clip_Multispectral_ L5015033_03320060504_MTL and sel_sheds.

3. Turn on the Image Classification toolbar and turn on the Spatial Analyst extension.

4. From the Classification drop-down menu, select Iso Cluster Unsupervised Classification. It automatically adds Clip_Multispectral_ L5015033_03320060504_MTL as the input raster.

Remember: This MTL file includes Landsat bands red, green, blue, NearInfared_1, NearInfrared_2, and MidInfrared.

5. Enter 40 as the number of classes. Accept the defaults for minimum class size and sample interval.

6. Name the file unsup and save it to your results folder. Click OK.

The Iso unsupervised classification scheme works by identifying clusters of pixels in the scene that have similar attributes. Determining the number of classes can be somewhat arbitrary, but you will typically make this number larger than the number of classes you want in your final display so that you will be able to clearly distinguish between clusters of different land-use types. Often, you will want to experiment with different values for the number of classes. The algorithm ignores clusters smaller than the minimum class size (so larger values for this parameter will lead to smaller clusters of pixels being ignored). The sample interval sets the spatial resolution of the classification. Lower values provide higher resolution but take considerably longer to process.

7. Go to Properties > Symbology and display unsup by unique values after running the tool and getting the results. It may take several seconds to a minute to complete the classification based on the scene and your computer. Upon completion, the final output will appear in the Table of Contents. Remove the Clip_Multispectral image.

8. Go to Properties > Symbology > Unique Values and display the raster by Value.

Q3 **What does each of the colors represent?**

Q4 **How does using more than three bands affect an unsupervised classification?**

Q5 **Compare the unsupervised classification image to the color composite image.**

MAKING
SPATIAL
DECISIONS
USING GIS
AND REMOTE
SENSING

UNSUPERVISED
CLASSIFICATION

DATA (Current Workspace)   \05unsupervised\project1_bay_unsup\bay_data_unsup
RESULTS (Scratch Workspace)   \05unsupervised\project1_bay_unsup\bay_results_unsup

107

STEP 3: Identify the classes

MAKING
SPATIAL
DECISIONS
USING GIS
AND REMOTE
SENSING

1

UNSUPERVISED
CLASSIFICATION

Now that the image has been classified, the task of identifying and merging classes remains. Your job is to classify the image into five classes: water, developed, forest, crop/pasture, and wetlands.

1. Add an imagery basemap using the Add Data button on the Standard toolbar.

2. Using the imagery basemap, try to identify the different land classes that were classified.

Hint: It is helpful to turn on the Effects toolbar and use the Swipe tool to swipe between the unsupervised classification and the imagery.

The classes of land you are trying to classify are (1) water, (2) developed, (3) forest, (4) crop/pasture, and (5) wetlands.

Use the table on your worksheet to record your findings.

Number	Class	Number	Class
1		21	
2		22	
3		23	
4		24	
5		25	
6		26	
7		27	
8		28	
9		29	
10		30	
11		31	
12		32	
13		33	
14		34	
15		35	
16		36	
17		37	
18		38	
19		39	
20		40	

Note: Water, agricultural land, and forest are land-cover categories that are easy to identify. However, the term *developed land* is very subjective. It can mean completely paved land, like an airport; partially paved land, like cities; or developments of houses (subdivisions). For this classification, developed land is defined as all of the above (completely paved, partially paved, and subdivisions).

Q6 *How well separated are the classes?*

Q7 *What classes have overlap that could cause confusion?*

STEP 4: Reclassify the classification

Next you are going to reclassify the unsupervised image into the five classes you identified in the table in step 3.

1. Use the Reclassify (Spatial Analyst) tool to reclassify unsup into the five classes of land. Using the values you established in the table, assign 1 to water, 2 to developed, 3 to forest, 4 to crop/pasture, and 5 to wetlands. When you open the Reclassify dialog box, click Unique to make the values appear separately. It is important that you are very careful when you reassign the New Values to the Old Values.

2. Use unsup as the input file. Name the output file rec_lc.

STEP 5: Perform postclassification processing

In the classified output, some isolated pixels or small regions may be misclassified. This gives the output a salt-and-pepper or speckled appearance. Postclassification processing removes the noise generated by these errors and improves the quality of the classified output. The Spatial Analyst toolbox provides a set of generalization tools for the postclassification processing task. For this project, the Majority Filter tool and the Boundary Clean tool will be used for postclassification processing.

Q8 *Describe the appearance of the classified image. Is it speckled? Does it have random pixels not assigned? Zoom in and examine the image closely.*

1. The Majority Filter tool removes isolated pixels from the classified image.
   a. Run the Majority Filter tool and set the input to unsup.
   b. Name the file unsupf and save it to your results folder.
   c. Accept the defaults.

2. Run the Majority Filter tool again.
   a. Set the input to unsupf.
   b. Name the file unsupf2 and save it to your results folder.

MAKING
SPATIAL
DECISIONS
USING GIS
AND REMOTE
SENSING

1

*UNSUPERVISED
CLASSIFICATION*

DATA (Current Workspace)   \05unsupervised\project1_bay_unsup\bay_data_unsup
RESULTS (Scratch Workspace)   \05unsupervised\project1_bay_unsup\bay_results_unsup

109

3. The Boundary Clean tool smooths the ragged class boundaries and clumps the classes.
   a. Run the Boundary Clean tool on the unsupf2 raster.
   b. Name the file unsupf2bc and save it to your results folder.
   c. Select Ascend as the sorting technique.

MAKING
SPATIAL
DECISIONS
USING GIS
AND REMOTE
SENSING

Ascend is a sorting technique that sorts zones in ascending order by size. Zones with smaller total areas have a higher priority to expand into zones with larger total areas as the boundaries are smoothed.

   d. Clear the "Run expansion and shrinking twice" check box. This option forces expansion and shrinking to run only once, according to the sorting type.

4. Remove unsupf and unsupf2.

5. Turn on the Effects toolbar. Use the Swipe tool to swipe between unsupf2bc and the original unsup.

Select several parts of the images and zoom in closely to see the difference between the original unsup and the postprocessed unsupf2bc.

Q9 **Describe the difference between the original unsup and the postprocessed unsupf2bc.**

Q10 **Discuss the pros and cons of postprocessing.**

6. Remove all files except unsupf2bc.

STEP 6: Label the classes

In this section, text fields will be added to your postprocessed data for clarity of identification and labeling.

1. Add a field to the attribute table for unsupf2bc. Hint: Use the Table Options drop-down menu.
   a. Name is Type.
   b. Type is Text.
   c. Length is 20.

2. Open the attribute table, turn on the Editor toolbar, and click Start Editing. Enter the class names in the field Type. Remember: 1 is assigned to water, 2 to developed, 3 to forest, 4 to crop/pasture, and 5 to wetlands.

3. Click Stop Editing and save your edits.

MAKING
SPATIAL
DECISIONS
USING GIS
AND REMOTE
SENSING

4. Make the land-cover types appropriate colors.

5. Display the land-cover types by unique values.

6. Save the land-cover types as a layer file to your results folder. Name the layer file landcover.

rec_lc

Rowid	VALUE	COUNT	TYPE
0	1	1104243	water
1	2	1682349	developed
2	3	834118	forest
3	4	2082698	crop/pasture
4	5	1563925	wetlands

STEP 7: Compare the three watersheds quantitatively

1. Select the Middle Potomac watershed.

2. Extract the data from unsupf2bc for the Middle Potomac watershed by using the Extract by Mask tool with the following parameters:
   a. Input raster is unsupf2bc.
   b. Input feature mask is sel_sheds with the Middle Potomac watershed selected.
   c. Name the output raster mid_lc and save it to your results folder.

3. Repeat directions 1 and 2 for the Patuxent and Severn watersheds. Name the output rasters pat_lc and sev_lc, respectively. Save the files to your results folder.

MAKING
SPATIAL
DECISIONS
USING GIS
AND REMOTE
SENSING

1

UNSUPERVISED
CLASSIFICATION

4. Clear all selections.

5. Go to Properties > Symbology and import the landcover layer file to each of the three watersheds.

6. Remove the unsupf2bc layer.

Look at each of the three watersheds individually to answer the following questions. You might want to add online imagery to help you.

Q11 **What is the dominant land cover of the Middle Potomac watershed? The Patuxent watershed? The Severn watershed?**

Q12 **Which watershed is nearest to the Chesapeake Bay? What is the dominant land cover of this watershed?**

Q13 **Which watershed has the highest imperviousness?**

Q14 **Which watershed(s) would have the greatest runoff from agriculture?**

Q15 **Can you identify areas where restoration could take place in each watershed?**

Q16 **What areas do you see that would be the most vulnerable and need the most protection?**

DATA (Current Workspace) \05unsupervised\project1_bay_unsup\bay_data_unsup
RESULTS (Scratch Workspace) \05unsupervised\project1_bay_unsup\bay_results_unsup

Deliverable 2: An unsupervised classification of the complete watershed area. The classes of land should include water, developed, forest, crop/pasture, and wetlands.

STEP 8: Graph the land cover in each watershed

MAKING
SPATIAL
DECISIONS
USING GIS
AND REMOTE
SENSING

1

UNSUPERVISED
CLASSIFICATION

In the previous questions, you qualitatively compared the watersheds based on observations. Although you can tell a good deal about the watershed land cover by observing the images, you can provide more detail if you add quantitative measurements.

In this section, you will calculate the percentages of the different types of land cover in each watershed. You can do this because you know both the total number of pixels in the watershed and the number of pixels of each land-cover type.

1. Open the attribute table of mid_lc.

2. Add a field to the mid_lc table.
    a. Name is Percent.
    b. Type is Float.

3. Right-click the field Count and click Statistics. From Statistics you can derive the total number of pixels in the watershed. Record the number of pixels to use in the following calculation.

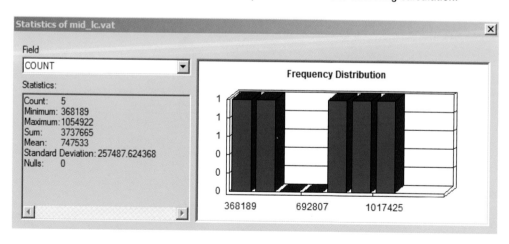

4. Right-click Percent and click Field Calculator. Enter the following formula to calculate the percentage of the different types of land cover.

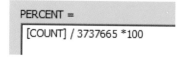

DATA (Current Workspace)   \05unsupervised\project1_bay_unsup\bay_data_unsup
RESULTS (Scratch Workspace)   \05unsupervised\project1_bay_unsup\bay_results_unsup

113

MAKING

SPATIAL

DECISIONS

USING GIS

AND REMOTE

SENSING

1

5. Repeat directions 1–4 for pat_lc and sev_lc.

Now you are ready to graph your findings.

6. Go to View > Graphs > Create Graph.

7. Create a graph for the Middle Potomac watershed with the following parameters:
   a. Vertical Bar.
   b. Value field is Percent.
   c. X label field is Type.
   d. Title the graph Percentage of Land Cover.

8. Add a series and graph the Patuxent watershed.

9. Add a series and graph the Severn watershed.

10. Right-click the top of the graph and add the graph to the layout.

11. Save as UnsupBay2.

Deliverable 3: A comparison graph with percentages of land cover for each of the three watersheds.

Deliverable 4: A written analysis using maps and graphs about the type of protection and restoration needed for each watershed.

MAKING
SPATIAL
DECISIONS
USING GIS
AND REMOTE
SENSING

2

*UNSUPERVISED
CLASSIFICATION*

PROJECT 2
Calculating unsupervised classification of Las Vegas, Nevada

Scenario/problem

A university in Nevada is doing a study of the variation in land cover of different watersheds in the area around Las Vegas. It has identified two watersheds for analysis: the Las Vegas Wash and the Detrital Wash. It has asked PtD to classify 2006 Landsat imagery into categories of land cover for the area (cf. Xian et al. 2008). It wants to use this land classification in a research proposal to request federal money for watershed restoration. It would like to analyze the watersheds based on the following land classes: evergreen, developed, shrub/scrub, and barren. It would like graphs showing percentages of these land types to draw conclusions from the analysis about the type of protection and restoration needed for each watershed.

Objectives

For this project, you will use Landsat imagery to do the following:
- Perform unsupervised classification with postprocessing cleanup of Landsat imagery.
- Combine the classes of the unsupervised classification into the categories of evergreen, developed, shrub/scrub, and barren.
- Compare the land cover of different watersheds.
- Use the classification to make recommendations for watershed protection and restoration.

DATA (Current Workspace)   \05unsupervised\project2_vegas_unsup\vegas_data_unsup
RESULTS (Scratch Workspace)   \05unsupervised\project2_vegas_unsup\vegas_results_unsup

MAKING
SPATIAL
DECISIONS
USING GIS
AND REMOTE
SENSING

2

UNSUPERVISED
CLASSIFICATION

Deliverables

We recommend the following deliverables for this project:

1. A map showing the Las Vegas Wash and the Detrital Wash watersheds with streams, places, and highways identified.
2. An unsupervised classification of each watershed with evergreen, developed, shrub/scrub, and barren labeled.
3. A comparison graph of percentages of land cover for the two watersheds.
4. A written analysis using maps and graphs about the type of protection and restoration needed for each watershed.

The questions for this project are both quantitative and qualitative. They identify key points that should be addressed in your analysis and presentation.

Keeping track of where your data and results are located is always a challenge. In these projects, we give the path to access the data (current workspace) and the path to store the results (scratch workspace) in a footnote on each page. The directions will specify whether the results go in the results folder or the results geodatabase.

Examine the data

This section was completed in module 1.

Organize and document your work

The data for this project is stored in the **\05unsupervised\project2_vegas_unsup\vegas_data_unsup** folder. Be sure to refer to project 1 and your process summary.

1. Set up the proper directory structure.

2. Create a process summary.

3. Document the map.

4. Set the environments:
    a. Data Frame Coordinate System to Projected Coordinate Systems > UTM > WGS 1984 > Northern Hemisphere > WGS 1984 UTM Zone 11N.
    b. Current Workspace to \05unsupervised\project2_vegas_unsup\vegas_data_unsup.
    c. Scratch Workspace to \05unsupervised\project2_vegas_unsup\vegas_results_unsup.
    d. Output Coordinate System to Same as Display.

Analysis

Once you have examined the data, completed the map documentation, and set the environments, you are ready to begin the analysis and complete the displays you need to address the problem. To begin this project, you need to identify and map the key watersheds.

MAKING
SPATIAL
DECISIONS
USING GIS
AND REMOTE
SENSING

2

UNSUPERVISED
CLASSIFICATION

STEP 1: Identify and map the designated watersheds

1. Start ArcMap and add L5039035_03520060512_MTL, watersheds, dtl_wat, and mjr_hwys from your data folder.

2. Select the Las Vegas Wash and Detrital Wash watersheds, extract by mask, and clip the various layers as needed.

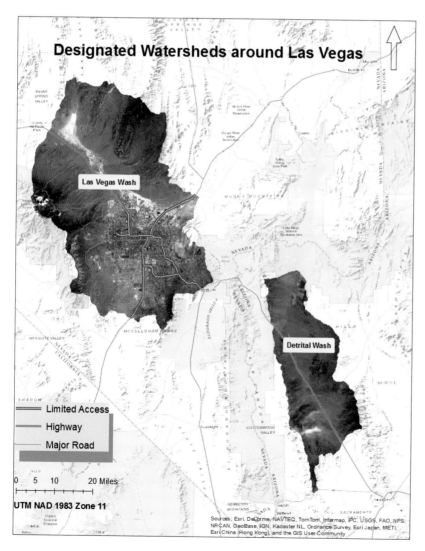

DATA (Current Workspace)   \05unsupervised\project2_vegas_unsup\vegas_data_unsup
RESULTS (Scratch Workspace) \05unsupervised\project2_vegas_unsup\vegas_results_unsup

117

MAKING
SPATIAL
DECISIONS
USING GIS
AND REMOTE
SENSING

2

UNSUPERVISED
CLASSIFICATION

Q1 *Add the Imagery with Labels basemap using the Add Data button on the Standard toolbar to help you identify the following features:*

   a. Lake Mead

   b. Lake Mohave

   c. Airport

   d. Canyons with vegetation

   e. Golf courses

   f. Mountains

   g. Patches of dark color appearing along the Vegas strip due to the shadows of tall buildings

Q2 *Try different band combinations: RGB_432, RGB_742, and RGB_453. Using the different band combinations, write a short analysis of the study area. Emphasize the land-cover patterns of the watershed.*

Deliverable 1: A map showing the Las Vegas Wash and Detrital Wash watersheds with streams, places, and highways identified.

Remember: Use the Swipe or Flicker tools from the Effects toolbar to compare the Landsat imagery to aerial imagery.

STEP 2: Classify Iso cluster unsupervised

Remember: Do not forget to turn on the Spatial Analyst extension.

1. Designate 20 **classes.**

2. Name the file and save it to your results folder.

Q3 *How does using more than three bands affect an unsupervised classification?*

Q4 *Compare the unsupervised image to the color composite image.*

Q5 *What does the remote sensing analyst control during the unsupervised classification process?*

DATA (Current Workspace)   \05unsupervised\project2_vegas_unsup\vegas_data_unsup
RESULTS (Scratch Workspace)   \05unsupervised\project2_vegas_unsup\vegas_results_unsup

STEP 3: Identify the classes

Q6   *Use the table on your worksheet to record your findings.*

Remember: The categories are evergreen, developed, shrub/scrub, and barren.

MAKING
SPATIAL
DECISIONS
USING GIS
AND REMOTE
SENSING

2

UNSUPERVISED
CLASSIFICATION

Number	Class	Number	Class
1		11	
2		12	
3		13	
4		14	
5		15	
6		16	
7		17	
8		18	
9		19	
10		20	

STEP 4: Reclassify the classification

1. Reclassify using the identified values for evergreen, developed, shrub/scrub, and barren.

STEP 5: Perform postclassification

1. Run Majority Filter twice.

2. Run Boundary Clean.

STEP 6: Label the classes

1. Make the land-cover types appropriate colors and label the types.

Deliverable 2: An unsupervised classification of each watershed with evergreen, developed, shrub/scrub, and barren labeled.

DATA (Current Workspace)   \05unsupervised\project2_vegas_unsup\vegas_data_unsup
RESULTS (Scratch Workspace)   \05unsupervised\project2_vegas_unsup\vegas_results_unsup

119

MAKING
SPATIAL
DECISIONS
USING GIS
AND REMOTE
SENSING

2

UNSUPERVISED
CLASSIFICATION

STEP 7: Quantitatively compare the two watersheds

1. Extract each watershed.

2. Calculate the percentage of each type of land cover.

Q7   *What is the dominant land use in each watershed?*

Q8   *What watershed has the highest imperviousness?*

Q9   *Can you identify vulnerable areas?*

STEP 8: Graph the land cover in each watershed

1. Calculate the percentage of each land-cover type in each watershed.

2. Graph the percentages comparing the land covers of each watershed. Because there are only two watersheds, create a pie graph of the percentages of land cover.

Deliverable 3: A comparison graph of percentages of land cover for the two watersheds.

Q10   *Compare and contrast the land cover of the Las Vegas watersheds with those of the Chesapeake Bay.*

     Las Vegas Wash watershed =

     Detrital Wash watershed =

Deliverable 4: A written analysis using maps and graphs about the type of protection and restoration needed for each watershed.

MAKING
SPATIAL
DECISIONS
USING GIS
AND REMOTE
SENSING

3

UNSUPERVISED
CLASSIFICATION

PROJECT 3
On your own

Scenario/problem

You have worked through a project performing an unsupervised classification of the Chesa-peake Bay region and repeated the analysis for the Las Vegas region. For this project, you will reinforce your skills by researching and analyzing a similar scenario entirely on your own. First, you must identify your study area and acquire the data for your analysis. You may want to study a local area. Refer to appendix A for directions on how to download your satellite imagery.

Research

Research the problem and answer the following questions:

1. What is the area of study?

2. What is the problem you are going to study?

3. What data is available?

Obtain the data

Do you have access to baseline data? Data and Maps for ArcGIS at http://www.esri.com/data/data-maps provides many of the layers of data that are needed for project work. Be sure to pay particular attention to the source of data and get the latest version. You can obtain data from the following sources:

MAKING
SPATIAL
DECISIONS
USING GIS
AND REMOTE
SENSING

3

UNSUPERVISED
CLASSIFICATION

- USGS Globalization Viewer at http://glovis.usgs.gov: Access to multiple sets of EROS satellite and aerial imagery.
- Census 2000 TIGER/Line Data at http://www.esri.com/tiger: Access to Census 2000 line data.
- Geospatial One-Stop at http://geo.data.gov: Web-based geospatial resources.
- The National Atlas at http://www.nationalatlas.gov: A range of products and geographic information about the United States.
- The National Map at http://nationalmap.gov/viewer.html: Data includes elevation, land cover, and topographic maps.

Workflow

After researching the problem and obtaining the data, you should do the following:

1. Write a brief scenario.

2. State the problem.

3. Define the deliverables.

4. Examine the metadata.

5. Set the directory structure, start your process summary, and document the map.

6. Decide what you need for the data frame coordinate system and the environments.
   a. What is the best projection for your work?
   b. Do you need to set a cell size or mask?

7. Start your analysis.

8. Prepare your presentation and deliverables.

9. Always remember to document your work in a process summary.

MODULE 6
SUPERVISED CLASSIFICATION

Introduction

In module 5, you used unsupervised classification to classify land cover in an image, essentially allowing ArcMap to use an algorithm to make classification choices. In this module, you will learn how to use supervised classification to do the same task yourself. Supervised classification gives the user much more input on how classification choices are made. You will also compare and contrast the two classification methods.

Scenarios in this module

- Calculating supervised classification of the Chesapeake Bay
- Calculating supervised classification of Las Vegas, Nevada
- On your own

Student worksheets

The student worksheet files can be found on the Maps and Data DVD.

Project 1: Chesapeake Bay student sheet
- File name: 06a_supervised_worksheet
- Location: \Student_Worksheets\06supervised

Project 2: Las Vegas student sheet
- File name: 06b_supervised_worksheet
- Location: \Student_Worksheets\06supervised

MAKING
SPATIAL
DECISIONS
USING GIS
AND REMOTE
SENSING

PROJECT 1
Calculating supervised classification of the Chesapeake Bay

Background

A major challenge in using Landsat imagery is to develop analysis schemes for using the data collected by the sensors at different wavelengths (spectral bands). One type of analysis involves categorizing the imagery into different land-cover types. There are different ways to do this categorization. In this project, you will focus on supervised classification. Supervised classification begins with collecting examples of the classes to appear in the final classified map. These are known as training samples, and they are used to obtain the spectral properties of the different features in the image.

Scenario/problem

After examination of the unsupervised classification of the three watersheds of the Chesapeake Bay analyzed in module 5, the Chesapeake Bay Foundation has asked for supervised classification of the watersheds. It would like to analyze the watersheds based on the same land-cover classes: water, developed, forest, crop/pasture, and wetlands; however, it would like two types of supervised classification to compare with the previous unsupervised classification. It would like an interactive supervised classification based on training samples and a maximum likelihood classification using spectral signatures. It would like graphs showing percentages of these land

DATA (Current Workspace)   \06supervised\project1_bay_sup\bay_data_sup
RESULTS (Scratch Workspace)   \06supervised\project1_bay_sup\bay_results_sup

types to analyze the type of protection and restoration needed for each watershed. It would also like a comparison with the unsupervised classification.

Objectives

For this project, you will use Landsat imagery to do the following:

- Perform an interactive supervised classification with training sites.
- Perform a maximum likelihood classification.
- Compare the percentages of land from the two supervised classifications with the unsupervised classification derived in module 5.

Deliverables

We recommend the following deliverables for this project:

1. An interactive supervised classification using training samples with five classes of land cover: water, developed, forest, crop/pasture, and wetlands.
2. A maximum likelihood supervised classification using spectral signatures with five classes of land cover: water, developed, forest, crop/pasture, and wetlands.
3. A comparison of percentages of land cover for both types of supervised classification and the unsupervised classification.

The questions for this project are both quantitative and qualitative. They identify key points that should be addressed in your analysis and presentation.

Keeping track of where your data and results are located is always a challenge. In these projects, we give the path to access the data (current workspace) and the path to store the results (scratch workspace) in a footnote on each page. The directions will specify whether the results go in the results folder or the results geodatabase.

Examine the data

This section was completed in module 1.

Organize and document your work

The following preliminary steps are essential to a successful geospatial analysis.

Examine the directory structure

In a geospatial project, you must carefully keep track of the data and your calculations. You will work with a number of different files, and it is important to keep them organized so you can

MAKING
SPATIAL
DECISIONS
USING GIS
AND REMOTE
SENSING

SUPERVISED
CLASSIFICATION

DATA (Current Workspace)   \06supervised\project1_bay_sup\bay_data_sup
RESULTS (Scratch Workspace)   \06supervised\project1_bay_sup\bay_results_sup

127

MAKING
SPATIAL
DECISIONS
USING GIS
AND REMOTE
SENSING

SUPERVISED
CLASSIFICATION

easily find them. The best way to do this is to have a folder for your project that contains a data folder. For this project, the folder named **\06supervised\project1_bay_sup\bay_data_sup** will be your project folder. Make sure it is stored in a place where you have write access.

You can store your data inside the results folder. The results folder already contains an empty geodatabase named **bay_results_sup** for this purpose. Save your map documents to the **bay_results_sup** folder.

Folder structure:

    06supervised
        project1_bay_sup
            bay_data_sup
                bay.gdb
                landsat_may_2006
            bay_results_sup
                bay_results_sup.gdb

Create a process summary

The process summary is simply a list of the steps you used to do your analysis. We suggest using a simple text document for your process summary. Keep adding to it as you do your work to avoid forgetting any steps. The following list shows an example of the first few entries in a process summary:

1. Create training samples for water, developed, forest, crop/pasture, and wetlands.
2. Investigate histograms of the training samples.
3. Merge training samples where needed.

Document the map

1. Start ArcMap and add descriptive properties to your map document properties.

2. Be sure to select the pathnames check box to store relative pathnames to all your data.

Set the environments

1. On the View menu, click Data Frame Properties. Set the map projection to Projected Coordinate Systems > UTM > WGS 1984 > Northern Hemisphere > WGS 1984 UTM Zone 18N.

2. Set the Current Workspace to \06supervised\project1_bay_sup\bay_data_sup.

3. Set the Scratch Workspace to \06supervised\project1_bay_sup\bay_results_sup.

4. For Output Coordinate System, select Same as Display.

5. Set the Processing Extent to the same as sel_sheds in the bay geodatabase and the Snap Raster to Multispectral_L5015015033_03320060504_MTL.

MAKING
SPATIAL
DECISIONS
USING GIS
AND REMOTE
SENSING

Analysis

Once you have examined the data, completed the map documentation, and set the environments, you are ready to begin the analysis and complete the displays you need to address the problem. For this project, you have been asked to perform supervised classifications.

STEP 1: Clip the multispectral Landsat scene to selected watershed boundaries

The Interactive Supervised Classification tool allows you to perform a supervised classification without explicitly creating a signature file. It makes use of all the bands available in the selected image. The user defines the training sites. A spectral signature file is simply a file that contains the spectral signatures (refer back to module 3) of different features in a scene that can be used to aid in supervised classification.

1. Add L5015033_03320060504_MTL from your data folder.

2. Go to Properties > Symbology and click Yes when asked if you want to make a histogram.

3. Reselect the band combination red, green, and blue by picking the channels again.

4. Add sel_sheds from your data folder. Use the Select by Rectangle tool to select all three watersheds.

5. On the Windows menu, click Image Analysis to open the window.

6. Select the MTL file, and then click the Clip tool. The Clip tool is the first tool in the Processing panel of the Image Analysis window.

7. This creates a temporary file called Clip_Multispectral_L5015033_03320060504_MTL.

8. Remove Multispectral_L5015033_03320060504_MTL from the Table of Contents.

You will be working with a lot of temporary files to do the supervised classification. You will not be required to save any of these files until you create your final product. However, because they are temporary files, you will need to finish the exercise in one session on your computer or else you will need to start over if you exit ArcMap.

MAKING
SPATIAL
DECISIONS
USING GIS
AND REMOTE
SENSING

1

SUPERVISED
CLASSIFICATION

9. Turn on the Image Classification toolbar.

The Image Classification toolbar (shown in the screen capture) provides a menu for creating both training samples and signature files.

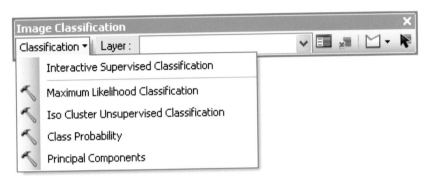

The following table shows the tools and buttons available on the Image Classification toolbar.

Button	Name	Function
	Training Sample Manager	Opens Training Sample Manager.
	Clear Training Samples	Removes all training samples and begins a new supervised classification session.
	Training Sample Drawing Tools	Draws new training samples in the display. There are three ways to create training samples: as polygons, as rectangles, and as circles.
	Select Training Sample	Selects training samples in the display.

STEP 2: Collect training samples

To the right of the Layer drop-down box you will see the Training Sample Manager. The Training Sample Manager allows you to manage training samples and name class values, merge classes, change display colors, and delete classes.

To the right of the Training Sample Manager are the drawing tools available for drawing polygons, circles, and rectangles. The following steps can be used to define a training sample:

1. Identify an area that belongs to a known class.
2. Draw a training sample using a drawing tool to enclose the area.
3. Once the training sample is drawn, a new class is created in the Training Sample Manager.

You have been tasked to determine five training samples: water, developed, forest, crop/pasture, and wetlands. Water is the easiest training sample, so you will create that first.

MAKING
SPATIAL
DECISIONS
USING GIS
AND REMOTE
SENSING

1

SUPERVISED
CLASSIFICATION

1. Zoom in and select several areas of water using the polygon drawing tool to create training samples. Try to get a variety of samples that represent the feature in different parts of the image.

2. Click all the training samples and select the Merge Training Samples feature tool directly above Class Name. Change the class name to Water and pick an appropriate color.

3. Follow the same procedure to make training samples for developed, forest, crop/pasture, and wetlands.

DATA (Current Workspace)   \06supervised\project1_bay_sup\bay_data_sup
RESULTS (Scratch Workspace)   \06supervised\project1_bay_sup\bay_results_sup

131

Wetlands are usually found near bodies of water. Forests are large green areas and are not necessarily adjacent to water. Crop/pasture land is usually green-brown in color and can be identified as near-rectangular shapes.

Be sure to collect enough signatures to represent the different types of land cover. Try to make your training sites as homogenous as possible (a single training site should have a single type of land cover).

MAKING
SPATIAL
DECISIONS
USING GIS
AND REMOTE
SENSING

SUPERVISED
CLASSIFICATION

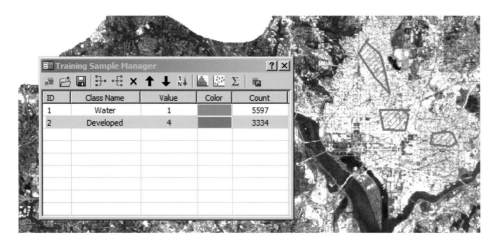

4. When you are done, your Training Sample Manager should look like the following image.

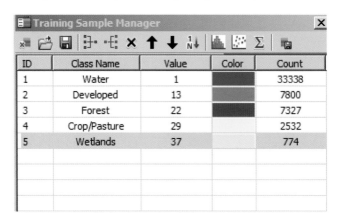

5. On the far right of the Training Sample Manager is a tool that allows you to create a signature file. Click the button and save the signature file to the bay_results_sup folder. Name the file signatures. You will use this file in a supervised classification in a later section under step 5.

STEP 3: Analyze the training samples

You now need to analyze the quality of your training samples. There are three tools that allow you to evaluate your training samples.

1. The Histograms window allows you to compare the distribution of the training samples. If the training samples represent different classes, their histograms should not overlap each other.

   Select all the training samples, and then click the Show histograms button. If you have more than four bands in the image layer, a vertical scrollbar will appear. The different colors of the histograms correspond to the training samples that represent your land-cover types. For example, blue represents the training samples for water, and green represents the training samples for forest.

Your histograms, scatterplots, and statistics will likely look different from the ones shown in the screen capture because they are dependent on the training samples you selected.

MAKING
SPATIAL
DECISIONS
USING GIS
AND REMOTE
SENSING

SUPERVISED
CLASSIFICATION

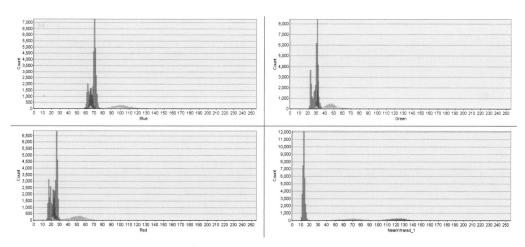

2. The Scatterplots window provides another way to compare multiple training samples. If the training samples represent different classes, their scatterplots should not overlap.

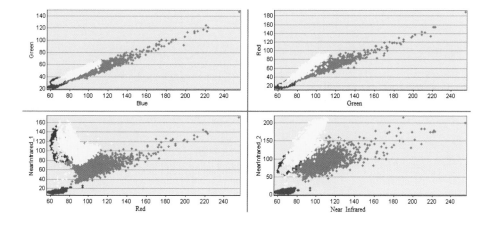

MAKING
SPATIAL
DECISIONS
USING GIS
AND REMOTE
SENSING

1

SUPERVISED
CLASSIFICATION

3. The Statistics window displays the statistics for the selected classes. Covariance indicates the extent to which the values in the different bands are correlated. Low values indicate separation of the classes in the two bands. For instance, the developed classification has high covariance in a number of band combinations, indicating that the classes are not well separated (as seen in the graph and histogram).

Water

Statistics	Blue	Green	Red	NearInfrared_1	NearInfrared_2	MidInfrared
Minimum	58.00	21.00	14.00	9.00	2.00	1.00
Maximum	94.00	43.00	44.00	26.00	19.00	11.00
Mean	68.31	28.67	23.04	12.97	8.09	5.17
Std.dev	3.59	3.11	3.87	1.33	1.93	1.32
Covariance						
Blue	12.87	10.04	11.80	3.10	0.97	0.47
Green	10.04	9.68	11.37	2.40	-0.17	-0.16
Red	11.80	11.37	14.95	2.93	-0.65	-0.43
NearInfrared_1	3.10	2.40	2.93	1.77	1.27	0.71
NearInfrared_2	0.97	-0.17	-0.65	1.27	3.73	1.84
MidInfrared	0.47	-0.16	-0.43	0.71	1.84	1.73

Developed

Statistics	Blue	Green	Red	NearInfrared_1	NearInfrared_2	MidInfrared
Minimum	63.00	29.00	21.00	30.00	36.00	18.00
Maximum	255.00	147.00	189.00	171.00	215.00	149.00
Mean	100.62	48.03	52.78	72.19	89.36	49.46
Std.dev	16.36	9.54	13.92	15.59	16.81	11.72
Covariance						
Blue	267.74	150.58	219.77	-27.30	159.25	145.19
Green	150.58	91.09	129.53	-0.10	107.18	88.00
Red	219.77	129.53	193.64	-15.58	153.82	132.31
NearInfrared_1	-27.30	-0.10	-15.58	243.15	111.44	-4.00
NearInfrared_2	159.25	107.18	153.82	111.44	282.46	168.32
MidInfrared	145.19	88.00	132.31	-4.00	168.32	137.30

4. Save your map document as supBay1 to your results folder.

Q1 *Which bands best differentiate between the four types of land?*

Q2 *Which bands have the greatest overlap, and what does this imply for using these bands in a classification?*

Q3 *From the graph can you tell whether the training sites represent the land cover accurately?*

DATA (Current Workspace)   \06supervised\project1_bay_sup\bay_data_sup
RESULTS (Scratch Workspace)   \06supervised\project1_bay_sup\bay_results_sup

STEP 4: Classify the image using an interactive supervised classification

A. Perform the classification

1. Close the windows that show the histograms, scatterplots, and statistics. Leave the Training Sample Manager and the Image Classification toolbar open.

2. You are now ready to classify the image. On the Classification drop-down menu on the Image Classification toolbar, click Interactive Supervised Classification. This creates a temporary file. Zoom out to the full extent of the image to see the full classification.

3. To make the file permanent, right-click the new layer and go to Data > Export Data.

   When you start to export the raster, it will ask whether you want to promote the pixel depth. Click Yes. This takes care of the situation where no NoData value is specified. So that you can retain the full range of values (256 in the case of an eight-bit raster), an extra value is added, forcing the raster to have 257 values and requiring it to be 16-bit (hence a "promotion" to a larger size). If there is a NoData value inside the input raster's data range or if there are NoData pixels, the pixel depth will not be changed.

   a. Save the file to your results folder.
   b. Name the file int_sup_bay and save it in GRID format.

4. Remove the temporary Classification_Clip_Multispectral file.

B. Conduct postprocessing tasks

In the classified output, some isolated pixels or small regions may be misclassified. This gives the output a salt-and-pepper or speckled appearance. Postclassification processing removes the noise generated by these errors and improves the quality of the classified output. The Spatial Analyst toolbox provides a set of generalization tools for the postclassification processing task. For this exercise, the Majority Filter and Boundary Clean tools will be used for postprocessing.

Q4 *Describe how the classified image looks. Is it speckled? Does it have random pixels not assigned?*

The Majority Filter tool removes the misclassified isolated pixels from the classified image.

MAKING
SPATIAL
DECISIONS
USING GIS
AND REMOTE
SENSING

SUPERVISED
CLASSIFICATION

DATA (Current Workspace)   \06supervised\project1_bay_sup\bay_data_sup
RESULTS (Scratch Workspace)   \06supervised\project1_bay_sup\bay_results_sup

135

MAKING
SPATIAL
DECISIONS
USING GIS
AND REMOTE
SENSING

1

SUPERVISED
CLASSIFICATION

1. Run the Majority Filter tool.
   a. Set the input to int_sup_bay.
   b. Name the output file intsupbayf and accept the defaults. Make sure the output file is saved to your results folder.

2. Run Majority Filter again.
   a. Set the input to intsupbayf.
   b. Name the output file intsupbayf2. **Make sure it is saved to your results folder.**

Boundary Clean smooths the ragged class boundaries and clumps the classes.

3. Run the Boundary Clean tool on the intsupbay2 raster.
   a. Name the output file intsupbayf2bc and save it to your results folder.
   b. Select Ascend for the sorting technique.
   c. Clear the "Run expansion and shrinking twice" check box.
   d. Remove int_sup_bay, intsupbayf, and intsupbayf2.

C. Label and create a layer file

1. Open the attribute table of intsupbayf2bc.

2. Add a field to the attribute table of intsupbayf2bc.
   a. Name is Type.
   b. Type is Text.
   c. Length is 20.

3. Open the attribute table, turn on the Editor toolbar, and click Start Editing. Enter the land-cover class names in the field Type. Remember: 1 is assigned to water, 2 to developed, 3 to forest, 4 to crop/pasture, and 5 to wetlands.

4. Stop editing and save your edits.

5. Symbolize the land cover by unique values. Select Type as the field.

6. Display the land-cover types with appropriate colors.

7. Save the symbology as a layer file to your results folder. Name the layer file landcover.

DATA (Current Workspace)   \06supervised\project1_bay_sup\bay_data_sup
RESULTS (Scratch Workspace)   \06supervised\project1_bay_sup\bay_results_sup

D. Quantitatively compare the three watersheds using interactive supervised classification

MAKING
SPATIAL
DECISIONS
USING GIS
AND REMOTE
SENSING

1. Select the Middle Potomac watershed.

2. Extract the data from intsupbayf2bc for the Middle Potomac watershed by using the Extract by Mask tool with the following parameters:
   a. Input raster is intsupbayf2bc.
   b. Input feature mask is sel_sheds with the Middle Potomac watershed selected.
   c. Name the output raster intsup_pot and save it to your results folder.

3. Repeat directions 1 and 2 for the Patuxent and Severn watersheds. Name the output rasters intsup_pat and intsup_sev, respectively. Save the files to your results folder.

4. Clear all selections.

5. Go to Properties > Symbology and import the landcover layer file to each of the three watersheds.

6. Remove the intsupbayf2bc layer.

E. Determine the percentages of land cover

1. Open the attribute table of intsup_pot.

2. Add a field.
   a. Name is Percent.
   b. Type is Float.
   c. Precision is 5.
   d. Scale is 2.

3. Right-click Count and click Statistics. Record the Sum of pixels.

4. Right-click Percent and click Field Calculator.

   The formula should be

   COUNT/3738993 * 100.

5. Repeat directions 1–4 for intsup_pat and intsup_sev.

DATA (Current Workspace)   \06supervised\project1_bay_sup\bay_data_sup
RESULTS (Scratch Workspace)   \06supervised\project1_bay_sup\bay_results_sup

137

6. Record the percentage of land cover in the chart on your worksheet.

7. Save your map document to your results folder.

Deliverable 1: An interactive supervised classification using training samples with five classes of land cover: water, developed, forest, crop/pasture, and forest.

MAKING
SPATIAL
DECISIONS
USING GIS
AND REMOTE
SENSING

SUPERVISED
CLASSIFICATION

STEP 5: Create a maximum likelihood classification based on spectral signatures

The maximum likelihood supervised classification uses the spectral signatures file you made and stored when you created the training samples. You named the file signatures and saved it to your results folder.

A. Perform the classification

1. Open the Image Classification toolbar, click Classification, and select Maximum Likelihood Classification.
   a. For Input raster bands, select Clip_Multispectral_L5015033_033006050_MTL.
   b. For Input signature file, select signatures from your results folder.
   c. Name the Output classified raster max_sup_bay and save it to your results folder.
   d. Click OK.

B. Perform postprocessing of the maximum likelihood classification

Repeat the postclassification processing directions from step 4.

The Majority Filter removes the misclassified isolated pixels from the classified image.

1. Run Majority Filter.
   a. Set the input to max_sup_bay and accept the defaults.
   b. Name the file maxsupbayf and save it to your results folder.

2. Run Majority Filter again.
   a. Set the input to maxsupbayf.
   b. Name the file maxsupbayf2 and save it to your results folder.

Boundary Clean smooths the ragged class boundaries and clumps the classes.

3. Run the Boundary Clean tool on the maxsupbayf2 raster.

    a. Select Ascend as the sorting technique.

    b. Clear the "Run expansion and shrinking twice" check box.

    c. Save the file as maxsupbayf2bc to your results folder.

    d. Remove maxsupbayf, maxupbayf2, and maxsupbayf2bc.

C. Label the field types

1. Open the attribute table of maxsupbayf2bc.

2. Add a field to the attribute table of maxsupbayf2bc.

    a. Name is Type.

    b. Type is Text.

    c. Length is 20.

3. Open the attribute table, turn on the Editor toolbar, and click Start Editing. Enter the class names in the field Type. Remember: 1 is assigned to water, 2 to developed, 3 to forest, 4 to crop/pasture, and 5 to wetlands.

4. Stop editing and save your edits.

5. Go to Properties > Symbology and import the layer file landcover from your results folder.

D. Calculate a quantitative comparison of the three watersheds using the maximum likelihood classification

1. Select the Middle Potomac watershed.

2. Extract the data from maxsupbayf2bc for the Middle Potomac watershed by using the Extract by Mask tool with the following parameters:

    a. Input raster is maxsupbayf2bc.

    b. Input feature mask is sel_states with the Middle Potomac watershed selected.

    c. Name the output raster maxsup_pot and save it to your results folder.

3. Repeat directions 1 and 2 for the Patuxent and Severn watersheds. Name the output rasters maxsup_pat and maxsup_sev. Save the files to your results folder.

4. Clear all selections.

MAKING
SPATIAL
DECISIONS
USING GIS
AND REMOTE
SENSING

*SUPERVISED
CLASSIFICATION*

DATA (Current Workspace)   \06supervised\project1_bay_sup\bay_data_sup
RESULTS (Scratch Workspace)   \06supervised\project1_bay_sup\bay_results_sup

139

MAKING
SPATIAL
DECISIONS
USING GIS
AND REMOTE
SENSING

1

*SUPERVISED
CLASSIFICATION*

5. Go to Properties > Symbology and import the landcover layer file to each of the three watersheds.

6. Remove the maxsupbayf2bc layer.

E. Determine the percentages of land cover

1. Open the attribute table of maxsup_pot.

2. Add a field.
   a. Name is Percent.
   b. Type is Float.
   c. Precision is 5.
   d. Scale is 2.

3. Right-click Count and click Statistics. Record the Sum of pixels:

   Pixels =

4. Right-click Count and click Field Calculator. The formula should be

   COUNT/3737816 * 100.

5. Repeat directions 1–4 for maxsup_pat and maxsup_sev.

6. Record the percentage of land cover in the chart on your worksheet.

7. Save your map document to your results folder.

Deliverable 2: A maximum likelihood supervised classification using spectral signatures with five classes of land cover: water, developed, forest, crop/pasture, and wetlands.

STEP 6: Create comparison graphs of percentages of land cover for both types of supervised classification and the unsupervised classification

You now need to analyze your findings. You have two different variables to deal with. The first variable is the difference in land cover between watersheds. The second variable is the three different types of classification that were performed: unsupervised, interactive supervised, and maximum likelihood supervised. The first step in the analysis is to get all the data into one place.

MAKING
SPATIAL
DECISIONS
USING GIS
AND REMOTE
SENSING

1. Enter the values in the chart on your worksheet. You must retrieve the data from module 5 for the unsupervised values.

PERCENTAGE OF LAND COVER BY CLASSIFICATION TYPE

	Middle Potomac			Patuxent			Severn		
	Unsupervised	Supervised		Unsupervised	Supervised		Unsupervised	Supervised	
		Interactive	Maximum likelihood		Interactive	Maximum likelihood		Interactive	Maximum likelihood
Water									
Developed									
Forest									
Crop/pasture									
Wetlands									

Deliverable 3: A comparison of percentages of land cover for both types of supervised classification and the unsupervised classification.

2. Analyze the data you have collected in the preceding chart and answer the following questions.

Q5   Compare and contrast the results of the unsupervised and supervised classifications for each watershed.

Q6   Compare and contrast the interactive supervised classification with the maximum likelihood supervised classification for each watershed.

Q7   Where would it be better to use an unsupervised classification?

Q8   Where would it be better to use a supervised classification?

Q9   In your opinion, which classification method produced the most accurate classification? Why?

MAKING
SPATIAL
DECISIONS
USING GIS
AND REMOTE
SENSING

1

SUPERVISED
CLASSIFICATION

Q10 *List two advantages and two disadvantages of using unsupervised classification.*

Q11 *List two advantages and two disadvantages of using supervised classification.*

Q12 *What is a mixed pixel? What effect does a mixed pixel have on classification techniques?*

DATA (Current Workspace)   \06supervised\project1_bay_sup\bay_data_sup
RESULTS (Scratch Workspace)   \06supervised\project1_bay_sup\bay_results_sup

MAKING
SPATIAL
DECISIONS
USING GIS
AND REMOTE
SENSING

2

SUPERVISED
CLASSIFICATION

PROJECT 2
Calculating supervised classification of Las Vegas, Nevada

Scenario/problem

After examination of the unsupervised classification of the two identified watersheds of the Las Vegas area produced in module 5, the urban planners have asked for a supervised classification of the watersheds. They would like PtD to analyze the two Las Vegas watersheds based on the same classes: evergreen, developed, shrub/scrub, and barren; however, they would like two types of supervised classification to compare with the previous unsupervised classification. They would like an interactive supervised classification based on training samples and a maximum likelihood supervised classification using spectral signatures. They would like graphs showing percentages of these land types to analyze the type of protection and restoration needed for each watershed. They would also like a comparison with the unsupervised classification.

Objectives

For this project, you will use Landsat imagery to do the following:
- Perform an interactive supervised classification with training sites.
- Perform a maximum likelihood supervised classification.
- Compare the percentages of land from the interactive and maximum likelihood classifications with the unsupervised classification derived in module 5.

DATA (Current Workspace)   \06supervised\project2_vegas_sup\vegas_data_sup
RESULTS (Scratch Workspace)   \06supervised\project2_vegas_sup\vegas_results_sup

143

MAKING
SPATIAL
DECISIONS
USING GIS
AND REMOTE
SENSING

2

*SUPERVISED
CLASSIFICATION*

Deliverables

We recommend the following deliverables for this project:

1. An interactive supervised classification using training samples with four classes of land cover: evergreen, developed, shrub/scrub, and barren.

2. A maximum likelihood supervised classification using spectral signatures with four classes of land cover: evergreen, developed, shrub/scrub, and barren.

3. A comparison of percentages of land cover for both types of supervised classification and the unsupervised classification.

The questions for this project are both quantitative and qualitative. They identify key points that should be addressed in your analysis and presentation.

Keeping track of where your data and results are located is always a challenge. In these projects, we give the path to access the data (current workspace) and the path to store the results (scratch workspace) in a footnote on each page. The directions will specify whether the results go in the results folder or the results geodatabase.

Examine the data

This section was completed in module 1.

Organize and document your work

The data for this project is stored in the **\06supervised\project2_vegas_sup\vegas_data_sup** folder. Be sure to refer to project 1 and your process summary.

1. Set up the proper directory structure.

2. Create a process summary.

3. Document the map.

4. Set the environments:
   a. Data Frame Coordinate System to Projected Coordinate Systems > UTM > WGS 1984 > Northern Hemisphere > WGS 1984 UTM Zone 11N.
   b. Current Workspace to \06supervised\project2_vegas_sup\vegas_data_sup.
   c. Scratch Workspace to \06supervised\project2_vegas_sup\vegas_results_sup.
   d. Output Coordinate System to Same as Display.

Analysis

Once you have examined the data, completed the map documentation, and set the environments, you are ready to begin the analysis and complete the displays you need to address the problem. To begin this project, you need to clip the Landsat scene to the watershed boundaries.

MAKING
SPATIAL
DECISIONS
USING GIS
AND REMOTE
SENSING

2

SUPERVISED
CLASSIFICATION

STEP 1: Clip the multispectral Landsat scene to selected watershed boundaries

1. Add L5039035_03520060512_MTL and watersheds from your data folder.

2. Isolate the Las Vegas Wash and Detrital Wash watersheds.

STEP 2: Collect training samples

1. Collect training samples for evergreen, developed, scrub/shrub, and barren.

2. Pick several examples of land cover for the training samples.

STEP 3: Analyze the training samples

1. Analyze training samples by histograms.

2. Analyze training samples by scatterplots.

3. Analyze training samples by statistics.

4. Calculate spectral signatures for future use.

Q1   *Which bands best differentiate between the four types of land?*

Q2   *Which bands have the greatest difference in the scatterplots and in what part of the spectrum?*

DATA (Current Workspace)   \06supervised\project2_vegas_sup\vegas_data_sup
RESULTS (Scratch Workspace)   \06supervised\project2_vegas_sup\vegas_results_sup

145

MAKING
SPATIAL
DECISIONS
USING GIS
AND REMOTE
SENSING

2

SUPERVISED
CLASSIFICATION

Q3   ***Which bands have the greatest overlap, and what does this imply for using these bands in a classification?***

Q4   ***From the graph can you tell whether the training sites represent the land cover correctly?***

STEP 4: Classify the image using an interactive supervised classification

A. Perform the classification

1. Create training samples.

2. Create spectral signatures.

3. Calculate an interactive supervised classification.

4. Remove temporary files.

B. Conduct postprocessing tasks

Q5   ***Describe how the classified image looks. Is it speckled? Does it have random pixels not assigned?***

1. Run Majority Filter twice.

2. Run Boundary Clean.

C. Label and create a layer file

1. Display land cover by appropriate color and name.

D. Quantitatively compare the three watersheds using interactive supervised classification

1. Extract the Las Vegas Wash watershed.

2. Extract the Detrital Wash watershed.

DATA (Current Workspace)   \06supervised\project2_vegas_sup\vegas_data_sup
RESULTS (Scratch Workspace)   \06supervised\project2_vegas_sup\vegas_results_sup

E. Determine the percentages of land cover

1. Calculate the percentage of the different land-cover types in each of the watersheds.

2. Enter the percentage results in the chart on your worksheet.

Deliverable 1: An interactive supervised classification using training samples with four classes of land cover: evergreen, developed, shrub/scrub, and barren.

MAKING
SPATIAL
DECISIONS
USING GIS
AND REMOTE
SENSING

STEP 5: Create a maximum likelihood classification based on spectral signatures

A. Perform the classification

1. Perform a maximum likelihood classification using the spectral signatures created when you made the training samples.

B. Perform postprocessing of the maximum likelihood classification

1. Run Majority Filter twice.

2. Run Boundary Clean.

C. Label the field types

1. Display the types of land cover by appropriate color and name.

D. Quantitatively compare the three watersheds using maximum likelihood supervised classification

1. Extract the Las Vegas Wash watershed.

2. Extract the Detrital Wash watershed.

E. Determine the percentages of land cover

1. Calculate the percentage of the different land-cover types in each of the watersheds.

2. Enter the percentage results in the chart on your worksheet.

Deliverable 2: A maximum likelihood supervised classification using spectral signatures with four classes of land cover: evergreen, developed, shrub/scrub, and barren.

MAKING
SPATIAL
DECISIONS
USING GIS
AND REMOTE
SENSING

2

SUPERVISED
CLASSIFICATION

STEP 6: Create comparison graphs of percentages of land cover for both types of supervised classification and the unsupervised classification

You are now faced with the job of trying to analyze all your findings. You have two different variables to analyze. The first variable is the difference in land cover between watersheds. The second is the three different types of classification that were performed: unsupervised, interactive supervised, and maximum likelihood supervised. The first step in the analysis is to get all the data into one place.

1. Enter the values in the following chart on your worksheet. You must retrieve the data from module 5 for the unsupervised values.

PERCENTAGE OF LAND COVER BY CLASSIFICATION TYPE

| | Las Vegas Wash | | | Detrital Wash | | |
| | Unsupervised | Supervised | | Unsupervised | Supervised | |
		Interactive	Maximum likelihood		Interactive	Maximum likelihood
Evergreen						
Developed						
Shrub/Scrub						
Barren						

Deliverable 3: A comparison of percentages of land cover for both types of supervised classification and the unsupervised classification.

2. Study the data you have collected in the preceding chart to answer the following questions.

Q6 *Compare and contrast the results of the unsupervised and supervised classifications for each watershed.*

Q7 *Compare and contrast the interactive supervised classification with the maximum likelihood supervised classification for each watershed.*

Q8 *Where would it be better to use an unsupervised classification?*

Q9 *Where would it be better to use a supervised classification?*

DATA (Current Workspace) \06supervised\project2_vegas_sup\vegas_data_sup
RESULTS (Scratch Workspace) \06supervised\project2_vegas_sup\vegas_results_sup

Q10 *In your opinion, which classification method produced the most accurate classification? Why?*

Q11 *List two advantages and two disadvantages of using unsupervised classification.*

Q12 *List two advantages and two disadvantages of using supervised classification.*

Q13 *What is a mixed pixel? What effect does a mixed pixel have on classification techniques?*

MAKING
SPATIAL
DECISIONS
USING GIS
AND REMOTE
SENSING

2

SUPERVISED
CLASSIFICATION

MAKING

SPATIAL

DECISIONS

USING GIS

AND REMOTE

SENSING

3

PROJECT 3
On your Own

Scenario/problem

You have worked through a project performing two supervised classifications on the Chesa-peake Bay region and repeated the analysis for the Las Vegas region. In this project, you will reinforce your skills by researching and analyzing a similar scenario entirely on your own. First, you must identify your study area and acquire the data for your analysis. You may want to do a local area. Refer to appendix A for directions on how to download your satellite imagery.

Research

Research the problem and answer the following questions:
1. What is the area of study?
2. What is the problem you are going to study?
3. What data is available?

Obtain the data

Do you have access to baseline data? Data and Maps for ArcGIS at **http://www.esri.com/data/data-maps** provides many of the layers of data that are needed for project work. Be sure to pay particular attention to the source of data and get the latest version. You can obtain data from the following sources:

- USGS Globalization Viewer at http://glovis.usgs.gov: Access to multiple sets of EROS satellite and aerial imagery.
- Census 2000 TIGER/Line Data at http://www.esri.com/tiger: Access to Census 2000 line data.
- Geospatial One-Stop at http://geo.data.gov: Web-based geospatial resources.
- The National Atlas at http://www.nationalatlas.gov: A range of products and geographic information about the United States.
- The National Map at http://nationalmap.gov/viewer.html: Data includes elevation, land cover, and topographic maps.

MAKING
SPATIAL
DECISIONS
USING GIS
AND REMOTE
SENSING

3

SUPERVISED
CLASSIFICATION

Workflow

After researching the problem and obtaining the data, you should do the following:

1. Write a brief scenario.

2. State the problem.

3. Define the deliverables.

4. Examine the metadata.

5. Set the directory structure, start your process summary, and document the map.

6. Decide what you need for the data frame coordinate system and the environments.
   a. What is the best projection for your work?
   b. Do you need to set a cell size or mask?

7. Start your analysis.

8. Prepare your presentation and deliverables.

9. Always remember to document your work in a process summary.

MODULE 7
CLASSIFICATION ACCURACY

Introduction

Once you have performed unsupervised and supervised classifications, it is important to determine the accuracy of your findings. In this module, you'll learn about errors of omission and commission, and then assess the accuracy of your classifications from modules 5 and 6.

Scenarios in this module

- Assessing the accuracy of land-cover data and different types of classification of the Chesapeake Bay
- Assessing the accuracy of land-cover data and different types of classification of Las Vegas, Nevada
- On your own

Student worksheets

The student worksheet files can be found on the Maps and Data DVD.

Project 1: Chesapeake Bay student sheet
- File name: 07a_accuracy_worksheet
- Location: \Student_Worksheets\07accuracy

Project 2: Las Vegas student sheet
- File name: 07b_accuracy_worksheet
- Location: \Student_Worksheets\07accuracy

MAKING
SPATIAL
DECISIONS
USING GIS
AND REMOTE
SENSING

PROJECT 1

1

Assessing the accuracy of land-cover data and different types of classification of the Chesapeake Bay

Background

To judge the reliability of a classified image, an assessment of its accuracy must be made. This accuracy assessment is generally based on ground referencing (sometimes called ground truthing). Ground referencing is the process of comparing data collected in the field or from higher spatial resolution imagery to a classified map derived from remotely sensed imagery. To ensure an accurate assessment of accuracy, your ground reference points must be taken at a scale that matches the imagery used for your classification (it wouldn't make sense to compare accuracy between 1-meter and 30-meter resolution images). Also, you need to have a sufficient number of random ground reference points for each land-cover class to provide reasonable statistics. Typically, 25–30 points per class would be a reasonable number.

Ground referencing of remotely sensed classified imagery is used to construct an error matrix, which is a table comparing the classes found on the ground or in the high spatial resolution imagery to those found in the classified image at the same location. This method uses errors of omission and errors of commission to compare the classified imagery to known points to assess

DATA (Current Workspace) \07accuracy\project1_bay_gt\bay_data_gt
RESULTS (Scratch Workspace) \07accuracy/project1_bay_gt\bay_results_gt

accuracy. Errors of omission are cases where categories have been omitted from the remotely sensed data that actually exist on the ground. Errors of commission are cases where identified categories in the remotely sensed data do not exist on the ground. To calculate errors of omission and commission, an error matrix is generated and ground-referenced values are used as the primary reference data.

MAKING
SPATIAL
DECISIONS
USING GIS
AND REMOTE
SENSING

1

*CLASSIFICATION
ACCURACY*

Scenario/problem

The Chesapeake Bay Foundation is requiring your company to do an accuracy assessment on the unsupervised and two supervised classifications in relation to both aerial imagery and the 2006 USGS land-cover data.

Objectives

For this project, you will use Landsat imagery to do the following:
- Generate random points and verify land cover from existing aerial photography at a spatial resolution of 30 meters.
- Calculate accuracy by establishing the reliability of the classifications comparing two maps: the ground-referenced map and the classified map.
- Generate error matrices to record data from ground truth maps and validate data from classified maps.
- Compare and contrast the accuracies of different classifications.

Deliverables

We recommend the following deliverables for this project:
1. A map of the multispectral imagery of the three designated watersheds with 25 random points, each with a 30-meter buffer. A table showing the 25 ground truth points generated from online imagery.
2. Omission and commission error matrices for the reclassified land cover 2006, reclassified unsupervised classification, reclassified interactive supervised classification, and reclassified maximum likelihood supervised classification.
3. A written analysis comparing the accuracy of the four classifications.

The questions for this project are both quantitative and qualitative. They identify key points that should be addressed in your analysis and presentation.

Keeping track of where your data and results are located is always a challenge. In these projects, we give the path to access the data (current workspace) and the path to store the results (scratch workspace) in a footnote on each page. The directions will specify whether the results go in the results folder or the results geodatabase.

DATA (Current Workspace)   \07accuracy\project1_bay_gt\bay_data_gt
RESULTS (Scratch Workspace)   \07accuracy/project1_bay_gt\bay_results_gt

155

Examine the data

This section was completed in module 1.

Organize and document your work

MAKING
SPATIAL
DECISIONS
USING GIS
AND REMOTE
SENSING

1

CLASSIFICATION
ACCURACY

The following preliminary steps are essential to a successful geospatial analysis.

Examine the directory structure

In a geospatial project, you must carefully keep track of the data and your calculations. You will work with a number of different files, and it is important to keep them organized so you can easily find them. The best way to do this is to have a folder for your project that contains a data folder. For this project, the folder named **\07accuracy\project1_bay_gt\bay_data_gt** will be your project folder. Make sure it is stored in a place where you have write access.

You can store your data inside the results folder. The results folder already contains an empty geodatabase named **bay_results_gt** for this purpose. Save your map documents to the **bay_results_gt** folder.

Folder structure:

    07accuracy
        project1_bay_gt
            bay_data_gt
                bay.gdb
                landsat_may_2006
            bay_results_gt
                bay_results_gt.gdb

Create a process summary

The process summary is simply a list of the steps you used to do your analysis. We suggest using a simple text document for your process summary. Keep adding to it as you do your work to avoid forgetting any steps. The following list shows an example of the first few entries in a process summary:

1. Extract by mask a multispectral image of designated watersheds.
2. Generate 25 random points.
3. Create 30-meter buffers around random points.

DATA (Current Workspace)   \07accuracy\project1_bay_gt\bay_data_gt
RESULTS (Scratch Workspace)   \07accuracy/project1_bay_gt\bay_results_gt

Document the map

1. Start ArcMap and add descriptive properties to your map document properties.

2. Be sure to select the pathnames check box to store relative pathnames to all your data.

MAKING
SPATIAL
DECISIONS
USING GIS
AND REMOTE
SENSING

1

CLASSIFICATION
ACCURACY

Set the environments

1. On the View menu, click Data Frame Properties. Set the map projection to Projected Coordinate Systems > UTM > WGS 1984 > Northern Hemisphere > WGS 1984 UTM Zone 18N.

2. Set the Current Workspace to \07accuracy\project1_bay_gt\bay_data_gt.

3. Set the Scratch Workspace to \07accuracy\project1_bay_gt\bay_results_gt.

4. For Output Coordinate System, select Same as Display.

5. Set the Processing Extent to the same as sel_sheds and the Snap Raster to Multispectral_ L5015015033_03320060504_MTL.

Analysis

Once you have examined the data, completed the map documentation, and set the environments, you are ready to begin the analysis and complete the displays you need to address the problem. To begin this project, you need to clip the Landsat scene to the watershed study area and generate random points.

STEP 1: Generate random points

1. Add L5015033_03320060504_MTL from your data folder.

2. Go to Properties > Symbology and click Yes when asked if you want to make a histogram.

3. Reselect the band combination red, green, and blue by selecting the channels again.

4. Add sel_sheds from your data folder and use the Select by Rectangle tool to select all three watersheds.

5. On the Windows menu, click Image Analysis to open the window.

DATA (Current Workspace)   \07accuracy\project1_bay_gt\bay_data_gt
RESULTS (Scratch Workspace)   \07accuracy/project1_bay_gt\bay_results_gt

157

MAKING
SPATIAL
DECISIONS
USING GIS
AND REMOTE
SENSING

CLASSIFICATION
ACCURACY

6. Select the MTL file, and then click the Clip tool. The Clip tool is the first button on the Processing panel of the Image Analysis window. This creates a temporary file called Clip_ Multispectral_L5015033_03320060504_MTL.

7. Remove the Multispectral_ L5015033_03320060504_MTL file. Clear selected features in sel_ sheds.

Before you create random points, you have to dissolve the three individual watersheds into one watershed.

8. On the main menu, go to Geoprocessing > Dissolve. Run the Dissolve tool:
   a. Input sel_sheds.
   b. Name the output feature class all_sheds and save it to your bay_results_gt geodatabase.
   c. Remove sel_sheds.

9. Use the Create Random Points tool (go to Data Management > Feature Class toolset).
   a. Output Location is your results folder.
   b. Output Point Feature Class is points_25.
   c. Constraining Feature Class is all_sheds.
   d. Number of Points is 25.
   e. Click OK.

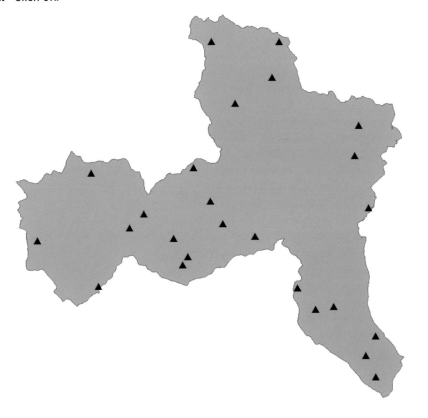

DATA (Current Workspace) \07accuracy\project1_bay_gt\bay_data_gt
RESULTS (Scratch Workspace) \07accuracy/project1_bay_gt\bay_results_gt

STEP 2: Create a 30-meter buffer around the random points

Landsat 5 imagery has a spatial resolution of 30 × 30 meters. Therefore, your sampling size for the ground reference points should be 30 meters. By creating 30-meter buffers around each point, you are defining a sampling area.

MAKING
SPATIAL
DECISIONS
USING GIS
AND REMOTE
SENSING

1. On the main menu, go to Geoprocessing > Buffer.
   a. Input feature is points_25.
   b. Output Feature Class is buff_25 and should be saved to your unsupbay_results geodatabase.
   c. The linear unit is 30.
   d. Click OK.

2. Zoom to a point and observe the 30-meter buffer zone.

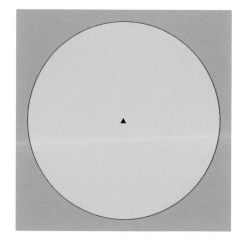

STEP 3: Ground truth buffered points from online imagery

1. Turn off the Clip_Multispectral image.

2. Turn off all_sheds.

3. Make the buff_25 layer hollow with an outline width of 2.

4. Add the Imagery basemap using the Add Data button on the Standard toolbar.

5. Open the points_25 attribute table and add a field:
   a. Name is GT.
   b. Type is Short Integer.
   c. Click OK.

DATA (Current Workspace)   \07accuracy\project1_bay_gt\bay_data_gt
RESULTS (Scratch Workspace)   \07accuracy/project1_bay_gt\bay_results_gt

159

Now comes the lengthy yet very important part of ground referencing. Zoom to each of the 25 points and make a judgment as to whether the land cover you are seeing is (1) water, (2) developed, (3) forest, (4) crop/pasture, or (5) wetlands.

Zoom in and out to get a more accurate feel for the land-cover type.

MAKING
SPATIAL
DECISIONS
USING GIS
AND REMOTE
SENSING

1

*CLASSIFICATION
ACCURACY*

6. Go to the Editor toolbar and click Start Editing. Enter the number that you identified for the land cover for each point in the attribute table.

  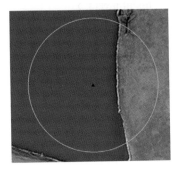

7. Click Stop Editing and save your edits.

8. Remove online imagery from the Table of Contents.

9. Remove Clip_Multispectral.

Deliverable 1: A map of the multispectral imagery of the three designated watersheds with 25 random points, each with a 30-meter buffer. A table showing the 25 ground reference points generated from online imagery.

STEP 4: Extract data from four different classification procedures

Now that you have identified the ground reference points, you are going to use them for an accuracy assessment of the following rasters:

- USGS Land Cover 2006: This land-cover classification is a product of the National Land Cover Database (NLCD). It has been created through a cooperative project conducted by the Multi-Resolution Land Characteristics Consortium (MRLC). The MRLC is a partnership of federal agencies, including the USGS, NOAA, and EPA. It is designed to provide updated land-cover data. The land-cover data was originally downloaded with all 20 classes defined in the NLCD. The original 2006 classification has been reclassified into the categories of (1) water, (2) developed, (3) forest, (4) crop/pasture, and (5) wetlands, which are the categories you are studying. It has been clipped to the sel_sheds feature class. The name of the file is lc_2006_rec.

DATA (Current Workspace)   \07accuracy\project1_bay_gt\bay_data_gt
RESULTS (Scratch Workspace)   \07accuracy/project1_bay_gt\bay_results_gt

- The previously derived unsupervised classification is called unsup from module 5.
- The derived interactive supervised classification is called int_sup from module 6.
- The derived maximum likelihood classification is called max_sup from module 6.

1. Add the files lc_2006_rec, unsup, int_sup, and max_sup.

2. Go to Properties > Symbology and import the landcover layer file from the unsupervised classification module (module 5) and the supervised classification module (module 6) for each of the raster layers.

MAKING
SPATIAL
DECISIONS
USING GIS
AND REMOTE
SENSING

CLASSIFICATION
ACCURACY

Q1 ***Turn the images on and off or use the Effects toolbar. What qualitative judgments can you make about the different classification schemes in relation to each other?***

Q2 ***Which classifications seem to be the most different from the others?***

Q3 ***Which classifications seem to be the most similar?***

You have already added a field GT to the feature class points_25. Now you must add the values from each of the different land-cover classifications to the points_25 feature class.

3. Search for and run the Extract Values to Points tool.
   a. Input point features is points_25.
   b. Input raster is lc_2006_rec.
   c. Save the file as points_25_2 to your results folder.
   d. Click OK.

When you open the attribute table of points_25_2, you will see that the lc_2006_rec values were added as the field RASTERVALU. Because you are using four rasters, you need to change the field name to keep track of which column is which.

   e. Add a short integer field to points_25_2. Name the field LC.
   f. Right-click the field LC and click Field Calculator.
   g. The formula LC = RASTERVALU will put all the values in the LC field.
   h. Right-click the RASTERVALU field and delete it.
   i. Remove points_25.

4. Repeat direction 3 for the input raster unsup.
   a. Name the file points_25_3.
   b. Add the field UNSUP.
   c. Remove the file points_25_2.

DATA (Current Workspace)   \07accuracy\project1_bay_gt\bay_data_gt
RESULTS (Scratch Workspace)   \07accuracy/project1_bay_gt\bay_results_gt

161

5. Repeat direction 3 for the raster int_sup.
   a. Name the file points_25_4.
   b. Add the field INT_SUP.
   c. Remove the file points_25_3.

6. Repeat direction 3 for the raster max_sup.
   a. Name the file points_25_5.
   b. Add the field MAX_SUP.
   c. Remove the file points_25_4.

MAKING
SPATIAL
DECISIONS
USING GIS
AND REMOTE
SENSING

1

CLASSIFICATION
ACCURACY

Your attribute table now contains not only the ground reference values but also the derived values from all the classification schemes. Having this data in one file allows you to proceed with the next step in accuracy analysis.

The data table you produce will likely vary from the table shown here because your reference points were assigned at random.

points_25_5

OID *	Shape *	CID	GT	LC	UNSUP	INT_SUP	MAX_SUP
1	Point	1	4	2	5	4	4
2	Point	1	1	1	1	1	1
3	Point	1	4	2	2	4	4
4	Point	1	3	3	4	3	3
5	Point	1	3	2	4	5	5
6	Point	1	3	3	5	3	3
7	Point	1	3	2	2	5	5
8	Point	1	1	1	1	1	1
9	Point	1	2	2	4	2	2
10	Point	1	4	2	5	4	4
11	Point	1	3	3	4	3	3
12	Point	1	4	2	5	4	4
13	Point	1	3	3	1	3	3
14	Point	1	3	2	3	5	5
15	Point	1	2	2	5	2	2
16	Point	1	3	3	4	5	5
17	Point	1	2	2	4	2	2
18	Point	1	2	2	2	2	2
19	Point	1	3	3	4	3	3
20	Point	1	3	3	5	3	3
21	Point	1	2	2	2	2	2
22	Point	1	1	3	1	2	2
23	Point	1	1	1	1	1	1
24	Point	1	2	4	2	4	4
25	Point	1	1	1	1	1	1

DATA (Current Workspace)   \07accuracy\project1_bay_gt\bay_data_gt
RESULTS (Scratch Workspace)   \07accuracy/project1_bay_gt\bay_results_gt

Create an error matrix for omission and commission

Two questions need to be answered when assessing accuracy:

- What percentage of the total classifications are correct (the overall accuracy)?
- What percentage of the classification of a given land-cover type is correct?

Calculating the percentage of correct classifications is simple. However, to determine the percentage of the classification of each type of land cover that is correct, you need to perform a more complicated calculation.

An error matrix that shows ground-referenced data in relation to classified data is a standard technique for this type of accuracy assessment. It allows you to calculate the total classification accuracy percentage as well as the accuracy percentage for individual land-cover types. The following table shows an error matrix to calculate the accuracy of the Reclassified USGS Land Cover 2006 image. (Microsoft Excel spreadsheets are provided for you to tabulate error matrices. They are found in the documents folder for this project: **\07accuracy\ project1_bay_gt\documents**.)

MAKING
SPATIAL
DECISIONS
USING GIS
AND REMOTE
SENSING

1

CLASSIFICATION
ACCURACY

Reclassified USGS Land Cover 2006

Classified category	Actual category: Ground truth					Total	User's accuracy (Errors of omission)
	(1) Water	(2) Developed	(3) Forest	(4) Agriculture	(5) Wetlands		
(1) Water							
(2) Developed							
(3) Forest							
(4) Agriculture							
(5) Wetlands							
Total							
Producer's accuracy (Errors of omission)							

The simplest statistic is overall accuracy, which is computed by dividing the total number of correct pixels (sum of the diagonal) by the total number of pixels in the error matrix.

The total number of correct pixels in a category divided by the total number of pixels of that category (column total) gives the omission error. Omission error occurs when we have omitted certain categories that actually exist on the ground. We say it isn't, and it is.

The total number of pixels in a category divided by the total number of pixels classified in that category (row) gives errors of commission. Errors of commission occur when we have identified categories that do not exist on the ground. We say it is, and it isn't.

DATA (Current Workspace)   \07accuracy\project1_bay_gt\bay_data_gt
RESULTS (Scratch Workspace)   \07accuracy/project1_bay_gt\bay_results_gt

163

MAKING
SPATIAL
DECISIONS
USING GIS
AND REMOTE
SENSING

1

CLASSIFICATION
ACCURACY

1. To complete the matrix, you need to use the Select By Attributes tool to identify the classifications that are correct. For example, to identify how many points were classified as 1 in the ground truth study and the LC study, use the following query, and then count the number of matches and enter it into the table. Repeat for each value. Your matrix is complete when you have matched all five types of land cover for this classification.

```
SELECT * FROM points_25_5 WHERE:
"GT" = 1 AND "LC" = 1
```

points_25_5

OID *	Shape *	CID	GT	LC	UNSUP	INT_SUP	MAX_SUP
1	Point	1	4	2	5	4	4
2	Point	1	1	1	1	1	1
3	Point	1	4	2	2	4	4
4	Point	1	3	3	4	3	3
5	Point	1	3	2	4	5	5
6	Point	1	3	3	5	3	3
7	Point	1	3	2	2	5	5
8	Point	1	1	1	1	1	1
9	Point	1	2	2	4	2	2
10	Point	1	4	2	5	4	4
11	Point	1	3	3	4	3	3
12	Point	1	4	2	5	4	4
13	Point	1	3	3	1	3	3
14	Point	1	3	2	3	5	5
15	Point	1	2	2	5	2	2
16	Point	1	3	3	4	5	5
17	Point	1	2	2	4	2	2
18	Point	1	2	2	2	2	2
19	Point	1	3	3	4	3	3
20	Point	1	3	3	5	3	3
21	Point	1	2	2	2	2	2
22	Point	1	1	3	1	2	2
23	Point	1	1	1	1	1	1
24	Point	1	2	4	2	4	4
25	Point	1	1	1	1	1	1

2. Complete the spreadsheet 07a_accuracy_bay in your documents folder. After entering the data, calculate the total percentage and errors of omission and commission.

Q4 *Which land-cover type has the highest error of commission?*

Q5 *Which land-cover type has the highest error of omission?*

Q6 *Which land-cover type is the most accurately classified?*

Q7 *Which land-cover type has the lowest classification accuracy?*

3. Repeat directions 1 and 2 for the unsupervised classification. Complete the corresponding matrix on your spreadsheet.

4. Repeat directions 1 and 2 for the interactive supervised classification. Complete the corresponding matrix on your spreadsheet.

5. Repeat directions 1 and 2 for the maximum likelihood supervised classification. Complete the corresponding matrix on your spreadsheet.

Q8 *Which type of classification is the most accurate for identifying forest land?*

Q9 *Which type of classification is the most accurate for identifying developed land?*

Deliverable 2: Omission and commission error matrices for the reclassified land cover 2006, reclassified unsupervised classification, reclassified interactive supervised classification, and reclassified maximum likelihood supervised classification.

Deliverable 3: A written analysis comparing the accuracy of the four types of classification.

MAKING
SPATIAL
DECISIONS
USING GIS
AND REMOTE
SENSING

CLASSIFICATION
ACCURACY

DATA (Current Workspace)   \07accuracy\project1_bay_gt\bay_data_gt
RESULTS (Scratch Workspace)   \07accuracy/project1_bay_gt\bay_results_gt

165

MAKING
SPATIAL
DECISIONS
USING GIS
AND REMOTE
SENSING

PROJECT 2
Assessing the accuracy of land-cover data and different types of classification of Las Vegas, Nevada

Scenario/problem

The urban planners are requiring PtD to do an accuracy assessment on its unsupervised and two supervised classifications in relation to aerial imagery and 2006 USGS land cover.

Objectives

For this project, you will use Landsat imagery to do the following:

- Generate random points and verify land cover from existing aerial photography at a spatial resolution of 30 meters.
- Quantify accuracy by establishing the reliability of the classifications by comparing two maps: the ground truth referenced map and the classified map.
- Generate error matrices to record data from ground truth maps and validate data from classified maps.
- Compare and contrast the accuracies of different classifications.

DATA (Current Workspace)   \07accuracy\project2_vegas_gt\vegas_data_gt
RESULTS (Scratch Workspace)   \07accuracy\project2_vegas_gt\vegas_results_gt

Deliverables

We recommend the following deliverables for this project:

1. A map of the multispectral imagery of the two designated watersheds with 25 random points and a 30-meter buffer. A table showing the ground truthing of 25 points using online imagery.
2. Omission and commission error matrices for the reclassified land cover 2006, reclassified unsupervised classification, reclassified interactive supervised classification, and reclassified maximum likelihood supervised classification.
3. A written analysis comparing the accuracy of the four types of classification.

The questions for this project are both quantitative and qualitative. They identify key points that should be addressed in your analysis and presentation.

Keeping track of where your data and results are located is always a challenge. In these projects, we give the path to access the data (current workspace) and the path to store the results (scratch workspace) in a footnote on each page. The directions will specify whether the results go in the results folder or the results geodatabase.

MAKING
SPATIAL
DECISIONS
USING GIS
AND REMOTE
SENSING

2

CLASSIFICATION
ACCURACY

Examine the data

This section was completed in module 1.

Organize and document your work

The data for this project is stored in the **\07accuracy\project2_vegas_gt\vegas_data_gt** folder. Be sure to refer to project 1 and your process summary.

1. Set up the proper directory structure.

2. Create a process summary.

3. Document the map.

4. Set the environments:
    a. Data Frame Coordinate System to Projected Coordinate Systems > UTM > WGS 1984 > Northern Hemisphere > WGS 1984 UTM Zone 18N.
    b. Current Workspace to \07accuracy\project2_vegas_gt\vegas_data_gt.
    c. Scratch Workspace to \07accuracy\project2_vegas_gt\vegas_results_gt
    d. Output Coordinate System to Same as Display.

DATA (Current Workspace)   \07accuracy\project2_vegas_gt\vegas_data_gt
RESULTS (Scratch Workspace)   \07accuracy\project2_vegas_gt\vegas_results_gt

167

Analysis

Once you have examined the data, completed the map documentation, and set the environments, you are ready to begin the analysis and complete the displays you need to address the problem. To begin this project, you need to clip the Landsat scene to the watershed study area and generate your random points.

MAKING
SPATIAL
DECISIONS
USING GIS
AND REMOTE
SENSING

2

CLASSIFICATION
ACCURACY

STEP 1: Extract the watershed study area and generate random points

1. Add L5039035_03520060512_MTL and sel_sheds from your data folder.

2. Create 24 random points and save them to your results folder.

The number of points should be 12. This will generate 12 points in each of the two watersheds.

STEP 2: Create a buffer around the random points

1. Create a 30-meter buffer around the random points and save it to your results folder.

STEP 3: Ground truth buffered points from online imagery

1. Add the Imagery basemap. Now comes the tedious part of ground truthing. Zoom to each of the 24 points and make a judgment call as to whether the land cover you are seeing is evergreen, developed, shrub/scrub, or barren. Zoom in and out to get a more accurate feel for the land-cover type.

2. Add a field to pts_24 and enter the ground truth results.

Deliverable 1: A map of the multispectral imagery of the two designated watersheds with 24 random points and a 30-meter buffer. A chart ground truthing the 24 points using online imagery.

STEP 4: Extract data from four different classification procedures

1. Add the 2006 land cover, unsupervised classification, interactive supervised classification, and maximum likelihood supervised classification.

Q1   **Turn the images on and off or use the Effects toolbar. What qualitative judgments can you make about the different classifications in relation to each other?**

Q2   **Which classifications seem to be the most different from the others?**

Q3   **Which classifications seem to be the most similar?**

You have already added a field GT to the feature class points_24. Now you must add the values from each of the different land-cover classifications to the points_24 feature class.

2. Run the Extract Values to Points tool for land cover and the unsupervised, interactive supervised, and maximum likelihood classifications.

MAKING
SPATIAL
DECISIONS
USING GIS
AND REMOTE
SENSING

2

CLASSIFICATION
ACCURACY

STEP 5: Create an error matrix for omission and commission

1. Complete the spreadsheet 07b_accuracy_vegas in your documents folder.

Q4   **Which land-cover type has the highest error of commission?**

Q5   **Which land-cover type has the highest error of omission?**

Q6   **Which land-cover type is the most accurately classified?**

Q7   **Which land-cover type has the lowest classification accuracy?**

Q8   **Which type of classification is the most accurate for identifying forest land?**

Q9   **Which type of classification is the most accurate for identifying developed land?**

Deliverable 2: Omission and commission error matrices for the reclassified land cover 2006, reclassified unsupervised classification, reclassified interactive supervised classification, and reclassified maximum likelihood supervised classification.

Deliverable 3: A written analysis comparing the accuracy of the four types of classification.

DATA (Current Workspace)   \07accuracy\project2_vegas_gt\vegas_data_gt
RESULTS (Scratch Workspace)   \07accuracy\project2_vegas_gt\vegas_results_gt

169

MAKING
SPATIAL
DECISIONS
USING GIS
AND REMOTE
SENSING

PROJECT 3
On your own

Scenario/problem

You have worked through a project assessing the accuracy of Chesapeake Bay region classifications and land cover and repeated the analysis for the Las Vegas region. For this project, you will reinforce your skills by researching and analyzing a similar scenario entirely on your own. First, you must identify your study area and acquire the data for your analysis. You may want to do a local area. Refer to appendix A for directions on how to download your satellite imagery.

Research

Research the problem and answer the following questions:

1. What is the area of study?
2. What is the problem you are going to study?
3. What data is available?

Obtain the data

Do you have access to baseline data? Data and Maps for ArcGIS at http://www.esri.com/data/data-maps provides many of the layers of data that are needed for project work. Be sure to pay particular attention to the source of data and get the latest version. You can obtain data from the following sources:

- USGS Globalization Viewer at http://glovis.usgs.gov: Access to multiple sets of EROS satellite and aerial imagery.

- Census 2000 TIGER/Line Data at http://www.esri.com/tiger: Access to Census 2000 line data.
- Geospatial One-Stop at http://geo.data.gov: Web-based geospatial resources.
- The National Atlas at http://www.nationalatlas.gov: A range of products and geographic information about the United States.
- The National Map at http://nationalmap.gov/viewer.html: Data includes elevation, land cover, and topographic maps.

MAKING
SPATIAL
DECISIONS
USING GIS
AND REMOTE
SENSING

3

CLASSIFICATION
ACCURACY

Workflow

After researching the problem and obtaining the data, you should do the following:

1. Write a brief scenario.

2. State the problem.

3. Define the deliverables.

4. Examine the metadata.

5. Set the directory structure, start your process summary, and document the map.

6. Decide what you need for the data frame coordinate system and the environments.
   a. What is the best projection for your work?
   b. Do you need to set a cell size or mask?

7. Start your analysis.

8. Prepare your presentation and deliverables.

9. Always remember to document your work in a process summary.

MODULE 8
URBAN CHANGE

Introduction

Remote sensing techniques are useful in a wide variety of situations, including the analysis of areas with rapid urban growth. In this module, you will use classification schemes to quantify the urban sprawl of two fast-growing locations: a Virginia county that is a suburb of Washington, DC, and Las Vegas, Nevada. Remote sensing techniques are very effective in assessing change over time.

Scenarios in this module

- Measuring urban change in Loudoun County, Virginia
- Measuring urban change in Las Vegas, Nevada
- On your own

Student worksheets

The student worksheet files can be found on the Maps and Data DVD.

Project 1: Loudoun County student sheet
- File name: 08a_urban_worksheet
- Location: \Student_Worksheets\08urban

Project 2: Las Vegas student sheet
- File name: 08b_urban_worksheet
- Location: \Student_Worksheets\08urban

MAKING
SPATIAL
DECISIONS
USING GIS
AND REMOTE
SENSING

1

URBAN CHANGE

PROJECT 1
Measuring urban change in Loudoun County, Virginia

Background

Monitoring urban sprawl and trying to identify both the pace and pattern of that spread has long been a concern of urban planners and policy makers. Urban sprawl means the loss of forest and agricultural land to houses, roads, and factories, which can lead to environmental challenges as well as changing demographics. On a regional scale, remote sensing satellite imagery is often used to monitor urban sprawl. Images from different time periods can be compared using supervised classification to show changes in the developed and undeveloped land.

Scenario/problem

The Chesapeake Bay Foundation is asking for a temporal study of Loudoun County, Virginia. Loudoun County is located in the northwest portion of the Washington, DC, metropolitan area. It is approximately 520 square miles in area. It is a county of diverse land use, with rapidly developing urban areas in the eastern portion and farmland and wooded areas in the western portion. It has been one of the fastest-growing counties in the United States since the late 1990s. The CBF would like an analysis of urban growth from 1990 to 2005.

DATA (Current Workspace)   \08urban\project1_loudoun_urban\loudoun_data_urban
RESULTS (Scratch Workspace)   \08urban\project1_loudoun_urban\loudoun_results_urban

Objectives

For this project, you will use Landsat imagery to do the following:

- Use an image service to acquire Landsat data from different years.
- Extract Landsat data from an image service for a study area.
- Create signatures for developed and other land-cover types.
- Perform interactive supervised classifications of the temporal datasets.
- Calculate and compare the change in amount of urban land during the prescribed time period.

MAKING
SPATIAL
DECISIONS
USING GIS
AND REMOTE
SENSING

Deliverables

We recommend the following deliverables for this project:

1. Three different multispectral datasets for the time period specified.
2. Training samples and interactive classifications for all three years.
3. Comparison of changes in urban land during the specified time period.

The questions for this project are both quantitative and qualitative. They identify key points that should be addressed in your analysis and presentation.

Keeping track of where your data and results are located is always a challenge. In these projects, we give the path to access the data (current workspace) and the path to store the results (scratch workspace) in a footnote on each page. The directions will specify whether the results go in the results folder or the results geodatabase.

Examine the data

This section was completed in module 1.

Organize and document your work

The following preliminary steps are essential to a successful geospatial analysis.

Examine the directory structure

In a geospatial project, you must carefully keep track of the data and your calculations. You will work with a number of different files, and it is important to keep them organized so you can easily find them. The best way to do this is to have a folder for your project that contains a data folder. For this project, the folder named **08urban\project1_loudoun_urban\loudoun_data_urban** will be your project folder. Make sure it is stored in a place where you have write access.

DATA (Current Workspace)   \08urban\project1_loudoun_urban\loudoun_data_urban
RESULTS (Scratch Workspace)   \08urban\project1_loudoun_urban\loudoun_results_urban

175

You can store your data inside the results folder. The results folder already contains an empty geodatabase named **loudoun_results_urban** for this purpose. Save your map documents to the **loudoun_results_urban** folder.

Folder structure:

08urban
    project1_loudoun_urban
        loudoun_data_urban
           bay.gdb
        loudoun_results_urban
           loudoun_results_urban.gdb

MAKING
SPATIAL
DECISIONS
USING GIS
AND REMOTE
SENSING

1

URBAN CHANGE

The process summary is simply a list of the steps you used to do your analysis. We suggest using a simple text document for your process summary. Keep adding to it as you do your work to avoid forgetting any steps. The following list shows an example of the first few entries in a process summary:

1. Connect to the image service.
2. Extract by Loudoun mask the three temporal Landsat datasets.
3. Create training samples for urban.

Note that an image service provides access to image data through a web-based service. The imagery.arcgisonline.com image service provides access to Landsat imagery created by the USGS. It provides access to Landsat images for a number of different time periods.

Document the map

1. Start ArcMap and add descriptive properties to your map document properties.

2. Be sure to select the pathnames check box to store relative pathnames to all your data.

Set the environments

1. On the View menu, select Data Frame Properties. Set the map projection to Projected Coordinate Systems > UTM > WGS 1984 > Northern Hemisphere > WGS 1984 UTM Zone 18N.

2. Set the Current Workspace to \08urban\project1_loudoun_urban\loudoun_data_urban.

3. Set the Scratch Workspace to \08urban\project1_loudoun_urban\loudoun_results_urban.

4. For Output Coordinate System, select Same as Display.

5. Save the map document as urban1 to your results folder.

DATA (Current Workspace) \08urban\project1_loudoun_urban\loudoun_data_urban
RESULTS (Scratch Workspace) \08urban\project1_loudoun_urban\loudoun_results_urban

Analysis

Once you have examined the data, completed the map documentation, and set the environments, you are ready to begin the analysis and complete the displays you need to measure urban change.

STEP 1: Obtain temporal multispectral imagery from an image server

In this section, you will access a web-based image service; extract Landsat images for 1990, 2000, and 2005; and export the images to a permanent file.

1. Add Loudoun from the bay geodatabase in your data folder.

2. On the Standard toolbar, click the Add Data button, and then click the Look in: drop-down arrow and select GIS Servers.

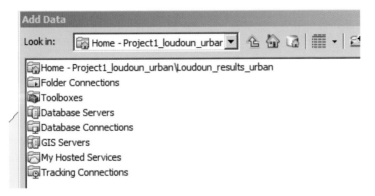

3. Click Add ArcGIS Server, and then click Add. In the next dialog box, click Use GIS Services, and then click Next.

4. Enter the server URL: http://imagery.arcgisonline.com/arcgis/services.

5. Click Finish.

MAKING
SPATIAL
DECISIONS
USING GIS
AND REMOTE
SENSING

URBAN CHANGE

DATA (Current Workspace)   \08urban\project1_loudoun_urban\loudoun_data_urban
RESULTS (Scratch Workspace)   \08urban\project1_loudoun_urban\loudoun_results_urban

177

MAKING

SPATIAL

DECISIONS

USING GIS

AND REMOTE

SENSING

1

URBAN CHANGE

6. Double-click arcgis on imagery.arcgisonline.com.

7. Double-click LandsatGLS.

8. Add TM_Multispectral_1990.

9. Select Loudoun using the Select by Rectangle tool.

10. Open the Image Analysis window.

11. At the top of the window, click LandsatGLS/TM_Multispectral_1990.

12. In the Processing panel, click Clip. This adds Clip_LandsatGLS\TM_Multispectral_1990 temporarily to your Table of Contents.

13. Remove LandsatGLS/TM_Multispectral_1990.

14. Run the Extract Band Function to make sure functions are using the proper bands. In the Image Analysis window, right-click the Clip_LandsatGLS\TM Multispectral_1990 image and click Properties. Then click the Functions tab.

15. When you click the Functions tab, the Raster Function Editor window is displayed. Right-click the Clip Function and click Insert to insert a function into the function chain. Click Extract Band Function. On the Extract Bands tab, for the Band input, enter the band numbers 1 2 3 4 5 6 one by one, with no commas in between, and click OK. This is to ensure that there is no confusion about which bands are used in subsequent processing functions.

16. To display enhanced Landsat imagery, in the Image Analysis window under Display, click the DRA option and select a standard deviation (std-dev) stretch.

17. Right-click the clipped temporary file in the Table of Contents and export it to your results folder. Name the file TM_1990 and save it as a TIFF image file to your results folder. When the "Export will add NoData pixels to the output raster dataset" warning appears, click Yes.

This image export process can take several minutes.

18. In the Image Analysis window, select TM_1990. Under Display, click the DRA option and select a standard deviation (std-dev) stretch.

19. Remove the temporary clipped file.

DATA (Current Workspace)   \08urban\project1_loudoun_urban\loudoun_data_urban
RESULTS (Scratch Workspace)   \08urban\project1_loudoun_urban\loudoun_results_urban

20. Repeat directions 9–19 for TM_Multispectral_2000. Save the file as TM_2000 to your results folder.

21. Repeat directions 9–19 for TM_Multispectral_2005. Save the file as TM_2005 to your results folder.

22. Save your map document to your results folder.

Turn the images on and off to observe the urban expansion.

Deliverable 1: Three different multispectral datasets for the time period specified.

Q1   ***Write a brief analysis describing the pattern of urban expansion in Loudoun County.***

STEP 2: Define training sites for developed and undeveloped land

For this analysis, the entire image will be the study area for Loudoun County.

1. Turn on the Image Classification toolbar.

2. Activate the Training Sample Manager. The next directions should be followed to define the training samples of developed and undeveloped land.

Because there is the least amount of developed land in the TM_1990 Landsat scene, you will use this scene to create the training sites that will be used for all three scenes. The success of your analysis depends on the quality of your training samples. For best results, choose only samples that you are certain represent developed land. For instance, do not confuse the high reflectance of a dirt field with the reflectance of concrete in an urban area. Second, be sure to include examples of all types of land that are considered to be undeveloped; for example, include fields, forest, and water in the undeveloped training samples.

  a. Select TM_1990 in the drop-down list.
  b. Identify an area that is developed.
  c. Draw a training sample using the Drawing tool to enclose the area.
   1. Draw several samples enclosing developed areas.
   2. Pick training samples of developed areas from the entire study area.
  d. Click all the training samples and select the Merge feature. Change the class name to developed and select an appropriate color.
  e. Follow directions a–d to make a training sample for undeveloped.

MAKING
SPATIAL
DECISIONS
USING GIS
AND REMOTE
SENSING

1

URBAN CHANGE

DATA (Current Workspace)   \08urban\project1_loudoun_urban\loudoun_data_urban
RESULTS (Scratch Workspace)   \08urban\project1_loudoun_urban\loudoun_results_urban

179

f. Click the Save button at the top of the Training Sample Manager and save your training samples to your results folder. Call the file samples. This gives you the option of loading the training samples again.

3. Save your map document to your results folder.

MAKING
SPATIAL
DECISIONS
USING GIS
AND REMOTE
SENSING

URBAN CHANGE

Deliverable 2: Training samples and interactive classifications for all three years.

STEP 3: Analyze the training samples

1. Select both the developed and undeveloped training sites.

2. Click the Show scatterplots button.

3. Examine the two training samples.

Q2 *Are the training samples distinct?*

Q3 *Do the training samples overlap in any of the bands?*

STEP 4: Perform an interactive supervised classification

You will now use the training samples to perform interactive supervised classifications of TM_1990, TM_2000, and TM_2005.

1. On the Image Classification toolbar, make sure TM_1990 is selected on the drop-down menu, and then click Interactive Supervised Classification on the Classification menu.

2. A temporary file named Classification_TM_1990 is created.

3. Repeat direction 1 for TM_2000 and TM_2005.

4. Right-click the names of each of the temporary files and export each one with file type Grid. Save the files to your results folder. Call the files TM_1990_GRID, TM_2000_GRID, **and** TM_2005_GRID. Click Yes to "Export will add NoData pixels to the output raster dataset."

5. Remove the temporary classification files.

DATA (Current Workspace)   \08urban\project1_loudoun_urban\loudoun_data_urban
       RESULTS (Scratch Workspace)   \08urban\project1_loudoun_urban\loudoun_results_urban

6. Qualitatively compare the urban growth in the three images. Different methods can be used to make this comparison:

a. Turn on the Effects toolbar and swipe the different temporal layers.

b. Make the undeveloped land hollow and turn the layers on and off.

Q4 **Is this visualization of the data more effective than looking at the false color composites? Why?**

MAKING
SPATIAL
DECISIONS
USING GIS
AND REMOTE
SENSING

STEP 5: Graph urban growth

Using the skills you have learned in previous modules, graph urban growth in Loudoun County from 1990 to 2005 using the value field Count.

Deliverable 3: Comparison of changes in urban land during the specified time period.

Q5 **You have been looking at the urban development in Loudoun County. How would you describe the land cover of the other areas of the county?**

Q6 **In the southeastern part of the county there has been massive urban development. What land cover has been lost due to this development? How could you show what land was transformed?**

Q7 **How would this type of analysis of urban growth help county officials in their planning to protect the environment, alleviate traffic congestion, and monitor groundwater availability?**

DATA (Current Workspace)   \08urban\project1_loudoun_urban\loudoun_data_urban
RESULTS (Scratch Workspace)   \08urban\project1_loudoun_urban\loudoun_results_urban

181

MAKING
SPATIAL
DECISIONS
USING GIS
AND REMOTE
SENSING

2

URBAN CHANGE

PROJECT 2
Measuring urban change in Las Vegas, Nevada

Scenario/problem

The urban planners are asking for a temporal study of Las Vegas, Nevada. Las Vegas is the largest city in the state of Nevada and is known for its gambling industry. Urban growth in Las Vegas has occurred in conjunction with railroads, the building of the Hoover Dam, and an influx of tourism and gambling. The rapid urban development has impacted an already fragile ecosystem. The urban planners want PtD to analyze urban growth from 1990 to 2005.

Objectives

For this project, you will use Landsat imagery to do the following:
- Use an image service to acquire different years of Landsat data.
- Extract Landsat data from an image service for a study area.
- Create signatures for developed and other land-cover types.
- Perform interactive supervised classifications on the temporal datasets.
- Calculate and compare the amount of urban land during the prescribed time period.

DATA (Current Workspace)   \08urban\project2_vegas_urban\vegas_data_urban
RESULTS (Scratch Workspace)   \08urban\project2_vegas_urban\vegas_results_urban

Deliverables

We recommend the following deliverables for this project:
1. Three different multispectral datasets for the time period specified.
2. Training samples and interactive classifications for the three dates.
3. Comparison of changes in urban land during the specified time period.

The questions for this project are both quantitative and qualitative. They identify key points that should be addressed in your analysis and presentation.

Keeping track of where your data and results are located is always a challenge. In these projects, we give the path to access the data (current workspace) and the path to store the results (scratch workspace) in a footnote on each page. The directions will specify whether the results go in the results folder or the results geodatabase.

MAKING
SPATIAL
DECISIONS
USING GIS
AND REMOTE
SENSING

2

URBAN CHANGE

Examine the data

This section was completed in module 1.

Organize and document your work

The data for this project is stored in the **08urban\project2_vegas_urban\vegas_data_urban** folder. Be sure to refer to project 1 and your process summary.

1. Set up the proper directory structure.

2. Create a process summary.

3. Document the map.

4. Set the environments:
   a. Data Frame Coordinate System to Projected Coordinate Systems > UTM > WGS 1984 > Northern Hemisphere > WGS 1984 UTM Zone 11N.
   b. Current Workspace to \08urban\project2_vegas_urban\vegas_data_urban.
   c. Scratch Workspace to \08urban\project2_vegas_urban\vegas_results_urban.
   d. Output Coordinate System to Same as Display.

DATA (Current Workspace)   \08urban\project2_vegas_urban\vegas_data_urban
RESULTS (Scratch Workspace)   \08urban\project2_vegas_urban\vegas_results_urban

183

Analysis

Once you have examined the data, completed the map documentation, and set the environments, you are ready to begin the analysis and complete the displays you need to measure urban change.

MAKING
SPATIAL
DECISIONS
USING GIS
AND REMOTE
SENSING

2

URBAN CHANGE

STEP 1: Obtain temporal multispectral imagery from an image server

1. Start ArcMap and add urban_SA from the vegas geodatabase in your data folder.

2. Access the GIS Server: http://imagery.arcgisonline.com/arcgis/services.

3. From the Landsat GLS, add TM_1990, TM_2000, and TM_2005.

4. Clip to and save the images with image type TIFF to your results folder.

Turn the images on and off to observe the urban expansion.

Deliverable 1: Three different multispectral datasets for the time period specified.

Q1   *Write a brief analysis explaining urban expansion in Las Vegas.*

STEP 2: Define training sites for developed and undeveloped land

1. Define training samples for developed and undeveloped land.

STEP 3: Analyze the training samples

1. Click the Show histograms button and examine the two training samples.

Q2   *Are the training samples distinct?*

Q3   *Do the training samples overlap in any of the bands?*

DATA (Current Workspace)   \08urban\project2_vegas_urban\vegas_data_urban
RESULTS (Scratch Workspace)   \08urban\project2_vegas_urban\vegas_results_urban

STEP 4: Perform an interactive supervised classification

1. Use the training samples to perform interactive supervised classifications for 1990, 2000, and 2005.

2. Use the Effects toolbar or turn the images on and off to observe the urban change.

3. Save the interactive supervised classifications as grids.

Q4   *Is this visualization of the data more effective than looking at the false color composites? Why?*

Deliverable 2: Training samples and interactive classifications for the three dates.

STEP 5: Graph urban growth

Using the skills you learned in previous modules, graph urban growth in Las Vegas from 1990 to 2005 using the value field Count.

Deliverable 3: Comparison of changes in urban land during the specified time period.

Q5   *You have been looking at the urban development in Las Vegas. How would you describe the land cover of the other areas of the surrounding county?*

Q6   *How would this type of analysis of urban growth help the urban planners in their long-range plans to protect the environment, alleviate traffic congestion, and monitor groundwater availability?*

MAKING
SPATIAL
DECISIONS
USING GIS
AND REMOTE
SENSING

2

URBAN CHANGE

DATA (Current Workspace)   \08urban\project2_vegas_urban\vegas_data_urban
RESULTS (Scratch Workspace)   \08urban\project2_vegas_urban\vegas_results_urban

185

MAKING
SPATIAL
DECISIONS
USING GIS
AND REMOTE
SENSING

3

URBAN CHANGE

PROJECT 3
On your own

Scenario/problem

You have worked through a project assessing urban change in Loudoun, Virginia, and repeated the urban change analysis for the Las Vegas region. In this project, you will reinforce your skills by researching and analyzing a similar scenario entirely on your own. First, you must identify your study area and acquire the data for your analysis. You may want to do a local area. Refer to appendix A for directions on how to download your satellite imagery.

Research

Research the problem and answer the following questions:

1. What is the area of study?
2. What is the problem you are going to study?
3. What data is available?

Obtain the data

Do you have access to baseline data? Data and Maps for ArcGIS at http://www.esri.com/data/data-maps provides many of the layers of data that are needed for project work. Be sure to pay particular attention to the source of data and get the latest version. You can obtain data from the following sources:

* USGS Globalization Viewer at http://glovis.usgs.gov: Access to multiple sets of EROS satellite and aerial imagery.

* Census 2000 TIGER/Line Data at http://www.esri.com/tiger: Access to Census 2000 line data.
* Geospatial One-Stop at http://geo.data.gov: Web-based geospatial resources.
* The National Atlas at http://www.nationalatlas.gov: A range of products and geographic information about the United States.
* The National Map at http://nationalmap.gov/viewer.html: Data includes elevation, land cover, and topographic maps.

MAKING
SPATIAL
DECISIONS
USING GIS
AND REMOTE
SENSING

Workflow

After researching the problem and obtaining the data, you should do the following:

1. Write a brief scenario.

2. State the problem.

3. Define the deliverables.

4. Examine the metadata.

5. Set the directory structure, start your process summary, and document the map.

6. Decide what you need for the data frame coordinate system and the environments.
   a. What is the best projection for your work?
   b. Do you need to set a cell size or mask?

7. Start your analysis.

8. Prepare your presentation and deliverables.

9. Always remember to document your work in a process summary.

MODULE 9
WATER

Introduction

Landsat images have an additional band that measures thermal radiation, band 6. This band allows you to compute the temperature of items in the scene. In this module, you will use band 6 to investigate the impact of drought on the size of lakes in Texas and calculate the temperature of lakes in Minnesota.

Scenarios in this module

- Measuring the impact of drought on Texas reservoirs
- Measuring Minnesota lake temperature using thermal infrared data
- On your own

Student worksheets

The student worksheet files can be found on the Maps and Data DVD.

Project 1: Texas student sheet
- File name: 09a_water_worksheet
- Location: \Student_Worksheets\09water

Project 2: Minnesota student sheet
- File name: 09b_water_worksheet
- Location: \Student_Worksheets\09water

MAKING
SPATIAL
DECISIONS
USING GIS
AND REMOTE
SENSING

1

WATER

PROJECT 1
Measuring the impact of drought on Texas reservoirs

Background

Remote sensing provides a means to map the spatial distribution of surface water. Landsat imagery has historically been used to monitor and measure area fluctuations in lakes. A six-year drought has caused the partial, and in some cases complete, evaporation of several reservoirs near San Angelo, Texas. Some of the reservoirs that used to be outdoor recreation areas are now gone, and the Texas Parks and Wildlife Department does not currently recommend the reservoirs as either fishing or boating destinations.

Scenario/problem

As a remote sensing specialist, you have been asked to show the depletion of the reservoirs by analyzing two Landsat scenes. One scene is from August 2009, and the other is from August 2011. You have been asked to analyze the shrinking area of the reservoirs both qualitatively, using multispectral imagery, and quantitatively, using the technique of density slicing.

DATA (Current Workspace)   \09water\project1_lake_fisher\lake_data_fisher
RESULTS (Scratch Workspace)   \09water\project1_lake_fisher\lake_results_fisher

Objectives

For this project, you will use Landsat imagery to do the following:
- Investigate individual Landsat bands to identify shrinking reservoir areas.
- Classify temporal Landsat data.
- Compare the sizes of the five reservoirs in the two scenes.

Deliverables

We recommend the following deliverables for this project:
1. An analysis of the study area with the designated reservoirs and Landsat imagery from 2009 and 2011, including a qualitative assessment of water area and quality based on your observations of the two Landsat scenes.
2. A density slice of 2009 and 2011 showing water and a quantitative comparison of the change in area of the reservoirs during this period.

The questions for this project are both quantitative and qualitative. They identify key points that should be addressed in your analysis and presentation.

Keeping track of where your data and results are located is always a challenge. In these projects, we give the path to access the data (current workspace) and the path to store the results (scratch workspace) in a footnote on each page. The directions will specify whether the results go in the results folder or the results geodatabase.

MAKING
SPATIAL
DECISIONS
USING GIS
AND REMOTE
SENSING

WATER

Examine the data

Q1 **View the item description and ArcGIS metadata for these features and complete the following chart on your worksheet.**

1. In ArcCatalog, right-click texas and view the Item Description.

2. In Windows Explorer, open the landsat\Aug_2_2009 folder and double-click the L5029038_03820090802_MTL text file to access the metadata of the Landsat image.

3. Complete the following chart on your worksheet.

Layer	Publication information: Who created the data?	Time period data is relevant	Spatial horizontal coordinate system	Data type	Resolution for rasters
texas					
L5029038_03820090802_B40					

DATA (Current Workspace)   \09water\project1_lake_fisher\lake_data_fisher
RESULTS (Scratch Workspace)   \09water\project1_lake_fisher\lake_results_fisher

191

Now that you have explored the available data, you are almost ready to begin your analysis. First you need to start a process summary, document your project, and set the project environments.

Organize and document your work

The following preliminary steps are essential to a successful geospatial analysis.

Examine the directory structure

In a geospatial project, you must carefully keep track of the data and your calculations. You will work with a number of different files, and it is important to keep them organized so you can easily find them. The best way to do this is to have a folder for your project that contains a data folder. For this project, the folder named **\09water\project1_lake_fisher\lake_data_fisher** will be your project folder. Make sure it is stored in a place where you have write access.

You can store your data inside the results folder. The results folder already contains an empty geodatabase named **fisher_results** for this purpose. Save your map documents to the **lake_results_fisher** folder.

Folder structure:
        09water
            project1_lake_fisher
                lake_data_fisher
                    landsat
                        Aug_2_2009
                        Aug_11_2011
                    texas.gdb
                lake_results_fisher
                    fisher_results.gdb

The process summary is simply a list of the steps you used to do your analysis. We suggest using a simple text document for your process summary. Keep adding to it as you do your work to avoid forgetting any steps. The following list shows an example of the first few entries in a process summary:

1. Make a basemap showing O. C. Fisher Lake and the surrounding area.
2. Make a layer file for lc_2006_sa.
3. Extract multispectral Landsat for 2009 and 2011 for the study area.

MAKING
SPATIAL
DECISIONS
USING GIS
AND REMOTE
SENSING

1

WATER

Document the map

1. Start ArcMap and add descriptive properties to your map document properties.

2. Be sure to select the pathnames check box to store relative pathnames to all your data.

MAKING
SPATIAL
DECISIONS
USING GIS
AND REMOTE
SENSING

Set the environments

1. On the View menu, click Data Frame Properties. Set the map projection to Projected Coordinate Systems > UTM > WGS 1984 > Northern Hemisphere > WGS 1984 UTM Zone 14N.

2. Set the Current Workspace to \09water\project1_lake_fisher\lake_data_fisher.

3. Set the Scratch Workspace to \09water\project1_lake_fisher\lake_results_fisher.

4. For Output Coordinate System, select Same as Display.

5. Save your map document as fisherlake1 to your results folder.

Analysis

Once you have examined the data, completed the map documentation, and set the environments, you are ready to begin the analysis and complete the displays you need to assess the impact of drought.

STEP 1: Observe Landsat scenes from 2009 and 2011

1. Add the layers texas, tx_counties, study_area, and reservoirs from the texas geodatabase.

Q2 **What are the three biggest reservoirs in the study area?**

2. Add L5029038_03820090802_MTL from landsat\Aug_2_2009 in your data folder.

Q3 **What do you think the dark area is in the middle of the study area?**

DATA (Current Workspace)   \09water\project1_lake_fisher\lake_data_fisher
RESULTS (Scratch Workspace)   \09water\project1_lake_fisher\lake_results_fisher

193

MAKING
SPATIAL
DECISIONS
USING GIS
AND REMOTE
SENSING

1

WATER

Q4   *Describe the water clarity and apparent depth in 2009 in each of the five reservoirs:*

>   *Oak Creek:*

>   *Ev Spence:*

>   *Lake Nasworthy:*

>   *Twin Buttes:*

>   *O. C. Fisher Lake:*

3. Change your false color composite to RGB_432 by resetting the red channel to NearInfrared_1, the green channel to Red, and the blue channel to Green.

Q5   *What is represented by the bright-red color in the southeastern corner of the scene?*

Q6   *When you zoom in on the image, you will see some perfect circles. What land cover is represented by these circles?*

4. Add L5029038_03820110808_MTL from landsat\Aug_11_2011 in your data folder.

Q7   *Describe the water clarity and apparent depth in 2011 in each of the five reservoirs:*

>   *Oak Creek:*

>   *Ev Spence:*

>   *Lake Nasworthy:*

>   *Twin Buttes:*

>   *O. C. Fisher Lake:*

5. Reset your false color composite to RGB_432 with the red channel set to NearInfrared_1, the green channel set to Red, and the blue channel set to Green.

Q8 *When you turn this image off and compare it with the 2009 image, what can you observe about the agricultural land to the southeast?*

Deliverable 1: An analysis of the study area with the designated reservoirs and Landsat imagery from 2009 and 2011. A written qualitative assessment of water area and quality based on your observations of the two Landsat scenes.

MAKING
SPATIAL
DECISIONS
USING GIS
AND REMOTE
SENSING

1

WATER

STEP 2: Assess USGS land cover for 2006

1. Add lc_2006_sa from your data folder.

2. Import the layer file lc_2006_sa.

Q9 *What is the dominant type of land cover?*

Q10 *What is the second-most dominant land cover?*

STEP 3: Define the band 4 study area

1. Add L5029038_03820090802_B40 from the Aug_2_2009 folder in your data folder.

2. Run the Extract by Mask tool with L5029038_03820090802_B40 as the input raster and study_area as the feature mask. Click Yes when asked if you want to promote the pixel depth. Name the output raster sabd42009 and save it to your results folder.

3. Go to Properties > Symbology and apply a Standard Deviations Stretch.

4. Remove L5029038_03820090802B40 from the Table of Contents.

5. Repeat directions 1–4 for L5029038_03820110808_B40 from the Aug_11_2011 folder. Name the output raster sabd42011.

DATA (Current Workspace)   \09water\project1_lake_fisher\lake_data_fisher
RESULTS (Scratch Workspace)   \09water\project1_lake_fisher\lake_results_fisher

195

STEP 4: Density slice the rasters

Slice is a raster tool that reclassifies the range of values of the input cells into zones of equal interval, equal area, or by natural breaks. The user designates the number of zones for reclassifying the input raster. In this project, you will designate three zones as you slice the two band 4 rasters (2009 and 2011). You will use natural breaks as the slice type. Natural breaks specifies that the classes are based on natural groupings observed in the data.

Because you are interested only in the reservoirs, you will use them as a mask. Setting an analysis mask means that processing will occur only on locations that fall within the mask, and all locations outside the mask will be assigned to NoData in the output. The Slice tool is located in the Spatial Analyst toolbox in the Reclass toolset.

1. Start the Slice tool:
    a. Input raster: sabd42009.
    b. Output raster: slice_2009.
    c. Number of zones: 3.
    d. Slice method: NATURAL_BREAKS.
    e. At the bottom of the Slice dialog box, click the Environments tab. In the Raster Analysis environment, set the Mask to reservoirs. Click OK.
    f. Click OK again.

2. Remove sabd42009.

3. Repeat directions 1 and 2 for L5029038_03820110808_B40 from the Aug_11_2011 folder. Name the output raster slice_2011.

Turn on 2009 and 2011 and examine your results. You will see that the deep water with some clarity is classified as 1, the turbid or cloudy water as 2, and the parts of the reservoir that look like wet vegetation as 3.

4. Create a layer file for slice_2009. Save the file as water to your results folder.

MAKING
SPATIAL
DECISIONS
USING GIS
AND REMOTE
SENSING

1

WATER

5. Import the layer file for slice_2011.

MAKING
SPATIAL
DECISIONS
USING GIS
AND REMOTE
SENSING

1

WATER

Q11 ***Looking at only the category of water, describe what happens to each of the reservoirs:***

    Oak Creek:

    Ev Spence:

    Lake Nasworthy:

    Twin Buttes:

    O. C. Fisher Lake:

DATA (Current Workspace)   \09water\project1_lake_fisher\lake_data_fisher
RESULTS (Scratch Workspace)   \09water\project1_lake_fisher\lake_results_fisher

197

STEP 5: Quantify your results

Because the water in the lake areas can be identified, the area can be calculated and the reduction in area of the lakes can be determined. The lakes need to be separated into individual rasters for meaningful calculations.

MAKING
SPATIAL
DECISIONS
USING GIS
AND REMOTE
SENSING

1

WATER

1. Select the Oak Creek Reservoir.

2. Extract by mask with slice_2009 as the input raster and reservoirs as the feature mask data. Name the file oak_2009 and save it to your results folder.

3. Repeat directions 1 and 2 for slice_2011. Name the file oak_2011.

Q12 *How many pixels represent deep water in oak_2009? In oak_2011?*

4. Repeat directions 1–3 for the Ev Spence reservoir.

Remember: You need to select the Ev Spence reservoir. Name the output files **ev_2009** and **ev_2011**.

2009.

DATA (Current Workspace)   \09water\project1_lake_fisher\lake_data_fisher
RESULTS (Scratch Workspace)   \09water\project1_lake_fisher\lake_results_fisher

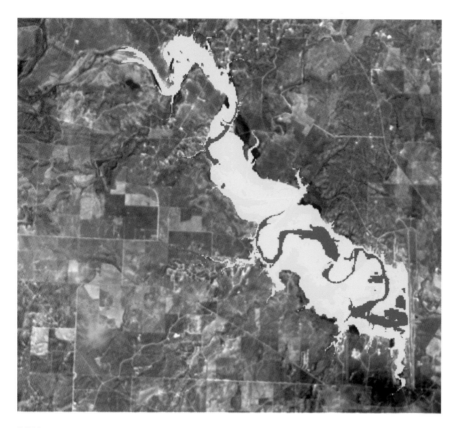

2011.

5. Repeat directions 1–3 for the Lake Nasworthy reservoir.

Remember: You need to select the Lake Nasworthy reservoir. Name the output files **nas_2009** and **nas_2011**.

6. Repeat directions 1–3 for the Twin Buttes Reservoir. Name the output files twin_2009 and twin_2011.

7. Repeat directions 1–3 for the O. C. Fisher Lake reservoir. Name the output files fish_2009 and fish_2011.

Q13 **Consolidate your findings in the following chart on your worksheet.**

	Pixels 2009	Pixels 2011	Shrinkage = pixels 2009 – pixels 2011	Percentage = (Shrinkage/pixels 2009) * 100
Oak Creek				
Ev Spence				
Lake Nasworthy				
Twin Buttes				
O.C. Fisher Lake				

MAKING
SPATIAL
DECISIONS
USING GIS
AND REMOTE
SENSING

1

WATER

MAKING

SPATIAL

DECISIONS

USING GIS

AND REMOTE

SENSING

1

WATER

Q14 *Which reservoir shrank the most?*

Q15 *Did any of the reservoirs expand?*

Deliverable 2: Density slice of 2009 and 2011 showing water and a quantitative comparison of the change in area of the reservoirs during this period.

DATA (Current Workspace)   \09water\project1_lake_fisher\lake_data_fisher
RESULTS (Scratch Workspace)   \09water\project1_lake_fisher\lake_results_fisher

MAKING
SPATIAL
DECISIONS
USING GIS
AND REMOTE
SENSING

PROJECT 2
Measuring Minnesota lake temperature using thermal infrared data

Scenario/problem

As a remote sensing specialist, you have been asked to compare the temperature of several lakes near Minneapolis, Minnesota. You have the thermal bands from two Landsat scenes. One scene is from June 1, 2009, and the other is from November 11, 2010. You have been asked to compare the thermal bands for the entire scene and to compare the temperatures of three specific lakes: Swede, Waconia, and Zumbra. Swede Lake is near Watertown, Minnesota; has a maximum depth of 12 feet; and covers 434 acres. Lake Waconia is near Waconia, Minnesota; has a maximum depth of 37 feet; and covers 3,080 acres. Zumbra Lake is near the town of Victoria, Minnesota; has a maximum depth of 58 feet; and covers 233 acres.

Objectives

For this project, you will use Landsat imagery to do the following:
- Investigate Landsat thermal band 6 comparing yearly temperature differences.
- Classify band 6 thermal data using a density slice.
- Compare the seasonal temperature differences of three lakes.

Deliverables

We recommend the following deliverables for this project:

1. A basemap showing Minneapolis, the study area, water, highways, and Swede, Waconia, and Zumbra Lakes identified.
2. An analysis of the Landsat thermal band 6 from June and November with quantitative data and graphs.
3. A density slice of Swede, Waconia, and Zumbra Lakes showing different thermal values for June and November.

The questions for this project are both quantitative and qualitative. They identify key points that should be addressed in your analysis and presentation.

Keeping track of where your data and results are located is always a challenge. In these projects, we give the path to access the data (current workspace) and the path to store the results (scratch workspace) in a footnote on each page. The directions will specify whether the results go in the results folder or the results geodatabase.

MAKING
SPATIAL
DECISIONS
USING GIS
AND REMOTE
SENSING

2

WATER

Examine the data

Q1 *View the item description and ArcGIS metadata for these features and complete the following chart on your worksheet.*

Layer	Publication information: Who created the data?	Time period data is relevant	Spatial horizontal coordinate system	Data type	Resolution for rasters
mn_wi					
L5027029_02920090601_B60					

Now that you have explored the available data, you are almost ready to begin your analysis. First you need to start a process summary, document your project, and set the project environments.

Organize and document your work

The data for this project is stored in the **\09water\project2_mn\mn_data** folder. Be sure to refer to project 1 and your process summary.

1. Set up the proper directory structure.

2. Create a process summary.

3. Document the map.

4. Set the environments.

Examine the directory structure

MAKING
SPATIAL
DECISIONS
USING GIS
AND REMOTE
SENSING

2

WATER

In a geospatial project, you must carefully keep track of the data and your calculations. You will work with a number of different files, and it is important to keep them organized so you can easily find them. The best way to do this is to have a folder for your project that contains a data folder. For this project, the folder named **\09water\project2_mn\mn_data** will be your project folder. Make sure it is stored in a place where you have write access.

You can store your data inside the results folder. The results folder already contains an empty geodatabase named **mn_results** for this purpose. Save your map documents to the **mn_results** folder.

Folder structure:
    09water
        project2_mn
            mn_data
                landsat
                    june_2009
                    nov_2010
                mn.gdb
            mn_results
                mn_results.gdb

The process summary is simply a list of the steps you used to do your analysis. We suggest using a simple text document for your process summary. Keep adding to it as you do your work to avoid forgetting any steps. The following list shows an example of the first few entries in a process summary:

1. Add thermal band 6 from your june_2009 folder.
2. Create an attribute table.
3. Create a line graph of DN thermal values.

Document the map

1. Start ArcMap and add descriptive properties to your map document properties.

2. Be sure to select the pathnames check box to store relative pathnames to all your data.

DATA (Current Workspace)   \09water\project2_mn\mn_data
RESULTS (Scratch Workspace)   \09water\project2_mn\mn_results

Set the environments

1. On the View menu, click Data Frame Properties. Set the map projection to Projected Coordinate Systems > UTM > WGS 1984 > Northern Hemisphere > WGS 1984 UTM Zone 15N.

MAKING
SPATIAL
DECISIONS
USING GIS
AND REMOTE
SENSING

2. Set the Current Workspace to \09water\project2_mn\mn_data.

3. Set the Scratch Workspace to \09water\project2_mn\mn_results.

4. For Output Coordinate System, select Same as Display.

5. Save your map document as mn_thermal to your results folder.

2

Analysis

An important first step in geospatial analysis is to create a basemap.

STEP 1: Create a basemap

1. Add the layers mn_wi, LG_study_area, mjr_highways, and water from the mn geodatabase.

2. Identify Swede, Waconia, and Zumbra Lakes.

3. Add L5027029_02920090601_MTL (June Landsat) and L5027029_02920101111_MTL (November Landsat). We recommend that you rename the two Landsat images so as not to confuse them with each other. We suggest June and Nov.

4. Create false color composites for June and November.

Q2   *Change the Landsat scenes to RGB_432. What type of land cover dominates the study area?*

Q3   *How is the June Landsat scene different from the November Landsat scene?*

Q4   *When you zoom to the study area, how does the water look different in the November scene?*

Q5   *In RGB_432, why is the June scene so much redder?*

Deliverable 1: A basemap showing Minneapolis, the study area, water, highways, and Swede, Waconia, and Zumbra Lakes identified.

DATA (Current Workspace)   \09water\project2_mn\mn_data
RESULTS (Scratch Workspace)   \09water\project2_mn\mn_results

STEP 2: Compare thermal bands from June and November

1. Add B60 from June and B60 from November.

2. Create a line graph comparing B60 June to B60 November.
   a. Go to View > Graphs > Create Graph.
   b. For Graph type, select Line Graph.
   c. For Value field, select Count.

MAKING
SPATIAL
DECISIONS
USING GIS
AND REMOTE
SENSING

2

WATER

Remember:

- A raster attribute table must be calculated for each band before it is viewed. Right-click each band and go to Properties > Symbology > Unique Values. The software will ask whether you want to build an attribute table. Click Yes and close the dialog box. Now the attribute table can be opened for the band. Change the classification back to Stretched in the Show window to change the view back to gray. Click OK.
- You do not want to graph the background values. Open the attribute table and select the value 0, which is the background value. Switch the selection to select the other values.

Q6 **What do the DN values represent in band 6?**

Q7 **What is the highest DN value in June band 6? The lowest? The mean?**

Q8 **What is the highest DN value in November band 6? The lowest? The mean?**

Q9 **Which month has the broadest range of DN values? Why?**

Q10 **Why are the DN values for November so concentrated?**

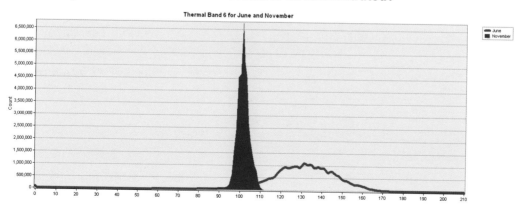

STEP 3: Convert DN to temperature

Band 6 is very useful for investigating the change in thermal properties of Landsat scenes, and the process for calculating temperature is very straightforward. However, you need to gather information from the metadata to calculate the gain and bias parameters.

MAKING
SPATIAL
DECISIONS
USING GIS
AND REMOTE
SENSING

2

WATER

In questions 6–10, DN values were used. These values can be converted to temperatures using the following procedure:

- Obtain the calibration parameters (LMIN, LMAX, QCALMAX, QCALMIN) for your specific Landsat image from the metadata. The metadata file for the June 2009 scene is \09water\project2_mn\mn_results\documents\L5027029_02920090601_MTL.txt. The calibration parameters are the same for band 6 in both the June and November Landsat scenes. Use these values to calculate the gain and bias for band 6.

To obtain a temperature from Landsat thermal band 6 data, you must convert the recorded DN value to radiance, and then calculate temperature from the radiance value based on the Planck function. Every sensor has some offset (bias) and gain (factor by which a measured quantity is attenuated to fit the range of the sensor). To convert the DN value to radiance, you need to know the characteristics of the Landsat thermal sensor: the offset or bias in the signal (LMIN), the range of possible DN values (QCALMAX – QCALMIN), and the range of possible radiance values that can be sensed (LMAX – LMIN). The gain is calculated using the following formula.

Do these calculations outside of ArcGIS:

Bias = LMIN

$$\text{Gain} = \frac{(\text{LMAX} - \text{LMIN})}{(\text{QCALMAX} - \text{QCALMIN})}$$

- Convert DN to radiance:

R = (G * DN) + B

where:

R = Pixel radiance value,

DN = Pixel digital number,

B = Bias (offset), and

G = Gain.

- Convert radiance value to temperature in degrees Kelvin:

$$T_k = \frac{K_2}{\ln\left(\frac{K_1}{R} + 1\right)}$$

DATA (Current Workspace)   \09water\project2_mn\mn_data
RESULTS (Scratch Workspace)   \09water\project2_mn\mn_results

where:

T_K = Temperature in degrees Kelvin,

R = Pixel radiance value,

K_1 = 607.76, and

K_2 = 1260.56.

- Convert T_k (Kelvin) to T_c (degrees Celsius)

T_c = T_k − 273.15.

MAKING
SPATIAL
DECISIONS
USING GIS
AND REMOTE
SENSING

2

WATER

Q11 *Add the feature class temperature. Record the DN at each of the three points for June band 6 and November band 6. Calculate the temperature in degrees Celsius and complete the following chart on your worksheet.*

Band 6 date	Point 1					Point 2					Point 3				
	DN	Radiance	K	°C	°F	DN	Radiance	K	°C	°F	DN	Radiance	K	°C	°F
June															
November															

Deliverable 2: An analysis of the Landsat thermal band 6 from June and November with quantitative data and graphs.

STEP 4: Density slice the rasters

1. Add the June thermal (junethermal) and November thermal (novthermal) rasters from the mn geodatabase. These are band 6 study areas extracted from the full scene.

2. Add lakes.

3. Run the Slice tool for both junethermal and novthermal:
   a. Use NATURAL_BREAKS.
   b. Use the selected lakes as a mask. Select both lakes and set the mask on the Environments tab at the bottom of the Slice dialog box.
   c. Number of zones: 3.

Turn on June and November and examine your results.

Q12 *Knowing what you do about the lakes, discuss the factors that would affect the heat they emit.*

Q13 *Why does there seem to be an edge around the lakes?*

Deliverable 3: A density slice of Swede, Waconia, and Zumbra Lakes showing different thermal values for June and November.

MAKING
SPATIAL
DECISIONS
USING GIS
AND REMOTE
SENSING

3

WATER

PROJECT 3
On your own

Scenario/problem

You have worked through a project assessing drought in Texas and a project assessing thermal energy in Minnesota lakes. For this project, you will reinforce your skills by researching and analyzing a similar scenario entirely on your own. First, you must identify your study area and acquire the data for your analysis. You may want to do a local area. Refer to appendix A for directions on how to download your satellite imagery.

Research

Research the problem and answer the following questions:
1. What is the area of study?
2. What is the problem you are going to study?
3. What data is available?

Obtain the data

Do you have access to baseline data? Data and Maps for ArcGIS at http://www.esri.com/data/data-maps provides many of the layers of data that are needed for project work. Be sure to pay particular attention to the source of data and get the latest version. You can obtain data from the following sources:
- USGS Globalization Viewer at http://glovis.usgs.gov: Access to multiple sets of EROS satellite and aerial imagery.

- Census 2000 TIGER/Line Data at http://www.esri.com/tiger: Access to Census 2000 line data.
- Geospatial One-Stop at http://geo.data.gov: Web-based geospatial resources.
- The National Atlas at http://www.nationalatlas.gov: A range of products and geographic information about the United States.
- The National Map at http://nationalmap.gov/viewer.html: Data includes elevation, land cover, and topographic maps.

MAKING
SPATIAL
DECISIONS
USING GIS
AND REMOTE
SENSING

3

WATER

Workflow

After researching the problem and obtaining the data, you should do the following:

1. Write a brief scenario.

2. State the problem.

3. Define the deliverables.

4. Examine the metadata.

5. Set the directory structure, start your process summary, and document the map.

6. Decide what you need for the data frame coordinate system and the environments.
   a. What is the best projection for your work?
   b. Do you need to set a cell size or mask?

7. Start your analysis.

8. Prepare your presentation and deliverables.

9. Always remember to document your work in a process summary.

MODULE 10
NORMALIZED DIFFERENCE VEGETATION INDEX

Introduction

Another important use of remote sensing is to assess the health of vegetation in different scenes. Satellite imagery can be used to create a measure of "greenness" called the Normalized Difference Vegetation Index (NDVI). In this module, you will use Landsat imagery to calculate NDVI and investigate drought and seasonal change.

Scenarios in this module

- Using NDVI to study drought conditions in Texas
- Using NDVI to study seasonal change in Minnesota
- On your own

Student worksheets

The student worksheet files can be found on the Maps and Data DVD.

Project 1: Texas student sheet
- File name: 10a_ndvi_worksheet
- Location: \Student_Worksheets\10ndvi

Project 2: Minnesota student sheet
- File name: 10b_ndvi_worksheet
- Location: \Student_Worksheets\10ndvi

MAKING
SPATIAL
DECISIONS
USING GIS
AND REMOTE
SENSING

1

PROJECT 1
Using NDVI to study drought conditions in Texas

Background

Many studies of drought that involve remote sensing use a vegetation index such as NDVI as an indicator of vegetation stress. A vegetation index is a combination of the reflectance at two different wavelengths and is used to study different properties of vegetation, such as vigor or greenness. The direct correlation between vegetation stress and NDVI has been established in remote sensing research (cf. Kogan 1997). NDVI has historically been used to monitor drought and make observations about agricultural land cover and is correlated to the greenness and relative biomass of vegetation.

NDVI is calculated using the difference of the spectral radiation of vegetation in the red and near-infrared bands. Healthy green leaves have a higher reflectivity in the near-infrared band. Stressed, dry, or diseased vegetation has much less reflectivity in the near-infrared band. NDVI is calculated using the following formula:

$$NDVI = (IR - R)/(IR + R)$$

where:

IR = the DN value of a particular pixel in the infrared band, and

R = the DN value of a particular pixel in the red band.

DATA (Current Workspace)   \10ndvi\project1_tx_ndvi\tx_data_ndvi
RESULTS (Scratch Workspace)   \10ndvi\project1_tx_ndvi\tx_results_ndvi

The preceding formula gives you a measure of greenness, with values between −1 and 1. Clouds and water have small negative values, rocks and barren land have small positive values, and healthy green vegetation has higher positive values.

Scenario/problem

Texas has been experiencing a severe drought situation for the past six years. Your consulting firm has been asked to quantify the severity of the drought by looking at NDVI of two complete Landsat scenes, one from 2009 and one from 2011. You have also been asked to compare NDVI of three different types of vegetation: crops, shrub/scrub, and evergreens. By comparing NDVI pre- and post-drought, you will be able to see which type of vegetation is most susceptible to drought conditions.

Objectives

For this project, you will use Landsat imagery to do the following:
- Use various methods to compare NDVI.
- Calculate NDVI for different Landsat scenes.
- Use graphs and statistics to compare temporal Landsat scenes.
- Use graphs and statistics to compare NDVI of different vegetation types.

Deliverables

We recommend the following deliverables for this project:
1. An analysis including NDVI images, graphs, and statistics comparing the Landsat scenes from 2009 and 2011.
2. Images isolating crops, shrub/scrub, and evergreen land cover.
3. Separate NDVI images, graphs, and statistics from 2009 and 2011 to compare crops, shrub/scrub, and evergreen land cover.

The questions for this project are both quantitative and qualitative. They identify key points that should be addressed in your analysis and presentation.

Keeping track of where your data and results are located is always a challenge. In these projects, we give the path to access the data (current workspace) and the path to store the results (scratch workspace) in a footnote on each page. The directions will specify whether the results go in the results folder or the results geodatabase.

MAKING
SPATIAL
DECISIONS
USING GIS
AND REMOTE
SENSING

1

NORMALIZED
DIFFERENCE
VEGETATION
INDEX

DATA (Current Workspace)   \10ndvi\project1_tx_ndvi\tx_data_ndvi
RESULTS (Scratch Workspace)   \10ndvi\project1_tx_ndvi\tx_results_ndvi

213

Examine the data

This section was completed in module 9.

Organize and document your work

MAKING
SPATIAL
DECISIONS
USING GIS
AND REMOTE
SENSING

1

NORMALIZED
DIFFERENCE
VEGETATION
INDEX

The following preliminary steps are essential to a successful geospatial analysis.

Examine the directory structure

In a geospatial project, you must carefully keep track of the data and your calculations. You will work with a number of different files, and it is important to keep them organized so you can easily find them. The best way to do this is to have a folder for your project that contains a data folder. For this project, the folder named **\10ndvi\project1_tx_ndvi\tx_data_ndvi** will be your project folder. Make sure it is stored in a place where you have write access.

You can store your data inside the results folder. The results folder already contains an empty geodatabase named **tx_results_ndvi** for this purpose. Save your map documents to the **tx_results_ndvi** folder.

Folder structure:
     10ndvi
          project1_tx_ndvi
               tx_data_ndvi
                    landsat
                         Aug_2_2009
                         Aug_11_2011
                    texas.gdb
               tx_results_ndvi
                    tx_results_ndvi.gdb

The process summary is simply a list of the steps you used to do your analysis. We suggest using a simple text document for your process summary. Keep adding to it as you do your work to avoid forgetting any steps. The following list shows an example of the first few entries in a process summary:
1. Add the two Landsat scenes 2009 and 2011.
2. Produce two NDVI images with a color map.
3. Create histogram graphs of NDVI.

Document the map

1. Start ArcMap and add descriptive properties to your map document properties.

DATA (Current Workspace)   \10ndvi\project1_tx_ndvi\tx_data_ndvi
RESULTS (Scratch Workspace)   \10ndvi\project1_tx_ndvi\tx_results_ndvi

2. Be sure to select the pathnames check box to store relative pathnames to all your data.

Set the environments

1. On the View menu, click Data Frame Properties. Set the map projection to Projected Coordinate Systems > UTM > WGS 1984 > Northern Hemisphere > WGS 1984 UTM Zone 14N.

2. Set the Current Workspace to \10ndvi\project1_tx_ndvi\tx_data_ndvi.

3. Set the Scratch Workspace to \10ndvi\project1_tx_ndvi\tx_results_ndvi.

4. For Output Coordinate System, select Same as Display.

5. Save your map document as ndvi1 to your results folder.

Analysis

Once you have examined the data, completed the map documentation, and set the environments, you are ready to begin the analysis and complete the displays you need to address the problem. A good place to start is to create a basemap of your study area.

STEP 1: Assess the drought of Texas

1. Add TX and tx_counties from your data folder.

2. Click the Add Data button and select Add Data From ArcGIS Online.

3. Search for Palmer Drought Severity index April–June 2011. This search will return multiple results. Select the one with the date 2/14/2011, and add this layer to your Table of Contents.

Q1   *What is the purpose of this data?*

4. Add L5029038_03820090802_MTL from Landsat\Aug_2_2009 in your data folder.

Q2   *What is the severity of drought in the area encompassed by the Landsat scene?*

5. Turn off the Palmer drought severity index forecast.

MAKING
SPATIAL
DECISIONS
USING GIS
AND REMOTE
SENSING

NORMALIZED
DIFFERENCE
VEGETATION
INDEX

DATA (Current Workspace)   \10ndvi\project1_tx_ndvi\tx_data_ndvi
RESULTS (Scratch Workspace)   \10ndvi\project1_tx_ndvi\tx_results_ndvi

215

STEP 2: Calculate NDVI using three different methods

A. Calculate NDVI by adding a function

MAKING
SPATIAL
DECISIONS
USING GIS
AND REMOTE
SENSING

1. Right-click the MTL Landsat scene and go to Properties > Functions > Function Chain. Right-click Composite Band Function and go to Insert > NDVI Function.

2. Set Input Raster to <Composite Band Function.OutputRaster>, Visible Band ID to 3, and Infrared Band ID to 4.

3. Click OK twice. A grayscale image appears.

You now need to add a Colormap function to produce an image that will be easier to interpret. This image is called a pseudocolor image, and the range of colors will represent different values of NDVI.

4. Return to the Functions tab on the Layer Properties dialog box. Right-click NDVI Function and go to Insert > Colormap Function. For Colormap, select NDVI. Click OK twice.

ArcGIS uses a slightly different formula to calculate and display NDVI. Using the formula below, NDVI values range from 0 to 200 (rather than −1 to 1).

$(IR − R)/(IR + R) * 100 + 100.$

Using this formula, NDVI values range from 0 to 200.

Q3 **What do the blue values represent?**

Q4 **Can you identify the red areas?**

Q5 **When you zoom in, some rectangular areas of darker green are visible, which is an indicator of high biomass. What do these areas of land cover represent?**

B. Calculate NDVI using the Image Analysis window with Colormap

Another way to calculate NDVI is to add the individual bands and use the Image Analysis window.

1. Add L5029038_03820090802_B30 and L5029038_03820090802_B40 from Landsat\Aug_2_2009 in your data folder. Add B30 first and B40 second.

2. Go to the Windows menu and activate the Image Analysis window.

3. In the upper panel, highlight both bands and click the NDVI button in the Processing panel. The NDVI icon is a leaf. By default, this NDVI operation adds two functions: the NDVI function and the Colormap function.

Q6   ***Compare the two different NDVI images: Are their values different? Do the color maps look different?***

MAKING
SPATIAL
DECISIONS
USING GIS
AND REMOTE
SENSING

NORMALIZED
DIFFERENCE
VEGETATION
INDEX

C. Calculate NDVI using the Image Analysis window with Scientific Output

You can also create NDVI using scientific output. This uses the Band Arithmetic function instead of the NDVI function. The Band Arithmetic function will output values between −1.0 and 1.0.

1. In the upper-left corner of the Image Analysis window, click the Options button.

2. Clear the Use Wavelength check box, and select the Scientific Output check box. Click OK.

3. Click B30 and B40 again, and then click the NDVI button.

You now see a grayscale image with values of −1.0 to 1.0. In this image, negative values are generated from areas with water, and values near zero are primarily generated from rock and bare soil. Very low values (0.1 and below) of NDVI correspond to barren areas of rock, and moderate values (0.2 to 0.3) represent shrub and grassland.

Q7   ***Discuss the advantages and disadvantages of each of the three methods used to generate NDVI images.***

4. Remove all layers except Texas and tx_counties.

5. Save your map document to your results folder.

DATA (Current Workspace)   \10ndvi\project1_tx_ndvi\tx_data_ndvi
RESULTS (Scratch Workspace)   \10ndvi\project1_tx_ndvi\tx_results_ndvi

217

STEP 3: Qualitatively compare NDVI from 2009 and 2011

MAKING
SPATIAL
DECISIONS
USING GIS
AND REMOTE
SENSING

After comparing the three methods of calculating NDVI, you need to choose a method for the following study. For consistency, choose the NDVI method with wavelengths that is calculated from the Image Analysis window. This is not necessarily the optimal method, but it will provide consistent results for comparison purposes. You have been provided with data that has been clipped to a study area, so the background noise has been eliminated.

1. Add band3_2009 and band4_2009 from your data folder.

2. Reposition band3_2009 above band4_2009 in the Table of Contents.

3. In the upper-left corner of the Image Analysis window, click the Options button. Clear the Scientific Output check box, and select the Use Wavelength check box. Click OK.

4. Highlight both bands, and then click the NDVI button in the Processing panel.

5. Remove band3_2009 and band4_2009 from the Table of Contents.

6. Repeat directions 1–5 using band3_2011 and band4_2011 from the Aug_11_2011 data folder.

Q8   ***Turn NDVI_band3_2011 off and on. Write a short summary comparing the two NDVI images.***

Q9   ***Looking at the reservoirs that you studied in module 9, does the 2011 NDVI image support your findings?***

7. Save your map document to your results folder.

STEP 4: Quantitatively compare NDVI from 2009 and 2011

In step 3, you could visually identify the difference in greenness between the 2009 and 2011 NDVI. You can quantitatively compare the two NDVI images using image statistics, and you can graph the NDVI data.

MAKING
SPATIAL
DECISIONS
USING GIS
AND REMOTE
SENSING

NORMALIZED
DIFFERENCE
VEGETATION
INDEX

1. Right-click each NDVI band and go to Properties > Functions > Function Chain and remove the color map.

2. Right-click NDVI_band3_2009 and go to Properties > Symbology, and then click Histograms. This gives you the image statistics data. The statistics calculated include the minimum and maximum pixel values as well as the mean and standard deviation of the calculated pixel values.

Q10 **Looking at the statistics for the 2009 and 2011 NDVI images, complete the following chart on your worksheet.**

The % change represents the percentage change of the NDVI values between 2009 and 2011.

	2009 NDVI	2011 NDVI	% Change
Minimum			
Maximum			
Mean			
Standard deviation			

Q11 **Use the statistics in the preceding chart to quantify your comparison analysis in question 8.**

Before you can graph the data, you need to convert the data to integer format.

3. Use the INT tool with NDVI_band3_2009 as the input raster.

4. Save the file as NDVI_2009 to your results folder.

5. Repeat directions 3 and 4 for NDVI_band3_2011 but save the file as NDVI_2011 to your results folder.

6. Graph NDVI_2009 and NDVI_2011 on the same axes.

DATA (Current Workspace)   \10ndvi\project1_tx_ndvi\tx_data_ndvi
RESULTS (Scratch Workspace)   \10ndvi\project1_tx_ndvi\tx_results_ndvi

219

You can connect the graph to the image by using the Select tool to select sections of the graph that will also be highlighted in the image.

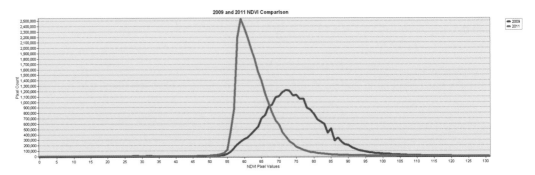

MAKING
SPATIAL
DECISIONS
USING GIS
AND REMOTE
SENSING

1

NORMALIZED
DIFFERENCE
VEGETATION
INDEX

Q12 ***Describe how the graph depicts the change in vegetation between 2009 and 2011.***

Deliverable 1: An analysis including NDVI images, graphs, and statistics comparing the Landsat scenes from 2009 and 2011.

STEP 5: Isolate vegetation types

1. Add lc_2006 from your data folder. This is from the USGS National Land Cover Database.

2. Go to Properties > Symbology and import the layer file landcover from your data folder.

Q13 ***What are the two most dominant land-cover types in the scene?***

Q14 ***You are concentrating on crops, shrub/scrub, and evergreen vegetation. Which is the least dominant land-cover type in the scene?***

3. Use the Raster Calculator to isolate shrub/scrub. The equation should read
   "lc_2006" == 52.

The Raster Calculator is a tool that allows you to create and execute Map Algebra expressions with raster data. The two equals signs (==) are part of the Raster Calculator syntax.

4. Name the output raster shrub and save it to your results folder.

5. The value of 1 represent shrubs. Make the 0 value hollow.

DATA (Current Workspace)   \10ndvi\project1_tx_ndvi\tx_data_ndvi
RESULTS (Scratch Workspace)   \10ndvi\project1_tx_ndvi\tx_results_ndvi

6. Repeat directions 3 and 4 for crops. The equation should read
    "lc_2006" == 82.

    Name the output raster crops and make the 0 value hollow.

7. Repeat directions 3 and 4 for evergreens. The equation should read
    "lc_2006" == 42.

    Name the output raster evergreens and make the 0 value hollow.

MAKING
SPATIAL
DECISIONS
USING GIS
AND REMOTE
SENSING

1

NORMALIZED
DIFFERENCE
VEGETATION
INDEX

The Times function in the Math toolset in the Spatial Analyst toolbox multiplies the values of two rasters on a pixel-by-pixel basis. If you multiply shrub by NDVI_2009, you will get a raster with only the NDVI values for shrubs. The other values will be 0, because the NDVI values have been multiplied by zero for the nonshrub pixels.

8. Run the Times tool and multiply shrub by NDVI_2009. Name the output raster shrub_2009.

9. Repeat direction 8 using NDVI_2011. Name the output raster shrub_2011.

10. Repeat directions 8 and 9 for crops. Name the output rasters crops_2009 and crops_2011.

11. Repeat directions 8 and 9 for evergreens. Name the files forest_2009 and forest_2011.

Deliverable 2: Images isolating crops, shrub/scrub, and evergreen land cover.

STEP 6: Compare NDVI of different vegetation types

You will now compare the statistics of NDVI for each vegetation type in 2009 and 2011. Right-click each of the six files and go to Properties > Symbology > Histograms to get the statistics for NDVI of each land-cover type.

Q15 *Enter the statistics in the following chart on your worksheet.*

	2009 NDVI values				2011 NDVI values				% change mean	% change mean
	min	max	mean	st dev	min	max	mean	st dev		
shrub										
crops										
evergreens										

DATA (Current Workspace)   \10ndvi\project1_tx_ndvi\tx_data_ndvi
RESULTS (Scratch Workspace)   \10ndvi\project1_tx_ndvi\tx_results_ndvi

221

1. Save your map document to your results folder.

Q16 *From the statistics in the preceding chart, which vegetation type was least affected by the drought?*

Q17 *What factors could make NDVI obtained for crops not be related to the drought?*

Deliverable 3: Separate NDVI images, graphs, and statistics from 2009 and 2011 to compare crops, shrub/scrub, and evergreen land cover.

MAKING
SPATIAL
DECISIONS
USING GIS
AND REMOTE
SENSING

NORMALIZED
DIFFERENCE
VEGETATION
INDEX

DATA (Current Workspace)   \10ndvi\project1_tx_ndvi\tx_data_ndvi
RESULTS (Scratch Workspace)   \10ndvi\project1_tx_ndvi\tx_results_ndvi

MAKING
SPATIAL
DECISIONS
USING GIS
AND REMOTE
SENSING

2

NORMALIZED
DIFFERENCE
VEGETATION
INDEX

PROJECT 2
Using NDVI to study seasonal change in Minnesota

Scenario/problem

Gopher State University's Remote Sensing and Analysis Laboratory has contracted you to provide a teaching/learning module explaining the value of understanding and using NDVI images. They have asked you to illustrate the module using Landsat data for a particular scene that focuses on the Minneapolis area where the university is located.

Objectives

For this project, you will use Landsat imagery to do the following:
- Calculate NDVI for different Landsat scenes.
- Use graphs and statistics to compare temporal Landsat scenes.
- Use graphs and statistics to compare NDVI of different vegetation types.

Deliverables

We recommend the following deliverables for this project:
1. An analysis including NDVI images, graphs, and statistics comparing the Landsat scenes from June 2009 and November 2010.
2. Images isolating crops, deciduous, and evergreen land cover.

DATA (Current Workspace)   \10ndvi\project2_mn_ndvi\mn_data_ndvi
RESULTS (Scratch Workspace)   \10ndvi\project2_mn_ndvi\mn_results_ndvi

223

3. Separate NDVI images, graphs, and statistics from June and November to compare crops, deciduous, and evergreen land cover.

The questions for this project are both quantitative and qualitative. They identify key points that should be addressed in your analysis and presentation.

MAKING
SPATIAL
DECISIONS
USING GIS
AND REMOTE
SENSING

Keeping track of where your data and results are located is always a challenge. In these projects, we give the path to access the data (current workspace) and the path to store the results (scratch workspace) in a footnote on each page. The directions will specify whether the results go in the results folder or the results geodatabase.

Examine the data

2

This section was completed in module 9.

Organize and document your work

The data for this project is stored in the **\10ndvi\project2_mn_ndvi\mn_data_ndvi** folder. Be sure to refer to project 1 and your process summary.

1. Set up the proper directory structure.

2. Create a process summary.

3. Document the map.

4. Set the environments:
   a. Data Frame Coordinate System to Projected Coordinate Systems > UTM > WGS 1984 > Northern Hemisphere > WGS 1984 UTM Zone 15N.
   b. Current Workspace to \10ndvi\project2_mn_ndvi\mn_data_ndvi.
   c. Scratch Workspace to \10ndvi\project2_mn_ndvi\mn_results_ndvi.
   d. Output Coordinate System to Same as Display.

Analysis

Once you have examined the data, completed the map documentation, and set the environments, you are ready to begin the analysis and complete the displays you need to calculate NDVI.

STEP 1: Compare NDVI from June and November

1. Add mn_wi from your data folder.

2. Add L5027029_02920090601_MTL from the Landsat\june_2009 data folder. Name the file June in the Table of Contents.

3. Add L5027029_02920101111_MTL from the Landsat\nov_2010 data folder. Name the file November in the Table of Contents.

Q1   ***Describe the difference between the two scenes.***

STEP 2: Calculate NDVI using the Image Analysis window with Colormap

1. Calculate NDVI for June.

2. Calculate NDVI for November.

Q2   ***What do the blue values represent?***

Q3   ***Can you identify the red/orange areas?***

Q4   ***When you zoom in, some symmetrical areas of darker green are visible, which is an indicator of high biomass. What do these areas of land cover represent?***

Q5   ***Compare June and November images: Are their values different? Do the color maps look different?***

STEP 3: Qualitatively compare NDVI in June and November

1. Remove the color map.

2. Calculate histograms and get the statistics.

MAKING
SPATIAL
DECISIONS
USING GIS
AND REMOTE
SENSING

2

NORMALIZED
DIFFERENCE
VEGETATION
INDEX

DATA (Current Workspace)   \10ndvi\project2_mn_ndvi\mn_data_ndvi
RESULTS (Scratch Workspace)   \10ndvi\project2_mn_ndvi\mn_results_ndvi

225

MAKING
SPATIAL
DECISIONS
USING GIS
AND REMOTE
SENSING

Q6 *Looking at the statistics for June NDVI and November NDVI, complete the following chart on your worksheet.*

The % change represents the percentage change of NDVI values between June and November.

	June NDVI	November NDVI	% Change
Mean			
Standard deviation			

Q7 *Use the statistics in the preceding chart to quantify your comparison analysis in question 5.*

3. Graph June and November NDVI.

Remember: To graph NDVI, you need to first process the NDVI rasters with the INT tool.

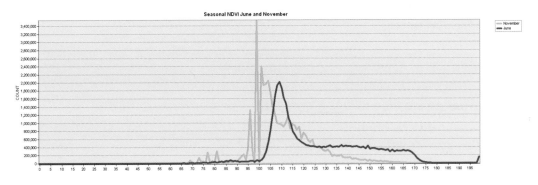

Q8 *Describe how the graph depicts the change in vegetation between June and November.*

Deliverable 1: An analysis including NDVI images, graphs, and statistics comparing the Landsat scenes from June 2009 and November 2010.

STEP 4: Isolate vegetation types

1. Add LC_2006 from your data folder.

Q9 *What are the two most dominant land-cover types in the scene?*

Q10 *You are concentrating on crops, deciduous, and evergreen vegetation. Which is the least dominant land-cover type in the scene?*

DATA (Current Workspace)   \10ndvi\project2_mn_ndvi\mn_data_ndvi
RESULTS (Scratch Workspace)   \10ndvi\project2_mn_ndvi\mn_results_ndvi

2. Isolate crops, deciduous, and evergreen vegetation. If you can't remember the numbers for different USGS land-cover types, refer to module 4.

3. Use the Times tool to combine the isolated vegetation and NDVI for June and November.

Deliverable 2: Images isolating crops, deciduous, and evergreen land cover.

STEP 6: Compare NDVI for different vegetation types

MAKING
SPATIAL
DECISIONS
USING GIS
AND REMOTE
SENSING

2

NORMALIZED
DIFFERENCE
VEGETATION
INDEX

You will now compare the statistics of NDVI for each vegetation type in June and November.

Q11 **Enter the statistics in the following chart on your worksheet.**

	June NDVI values		November NDVI values		% change mean	% change mean
	mean	st dev	mean	st dev		
crops						
deciduous						
evergreens						

Q12 **From the statistics in the preceding chart, which vegetation type changes the most with the season? The least?**

Deliverable 3: Separate NDVI images, graphs, and statistics from June and November to compare crops, deciduous, and evergreen land cover.

DATA (Current Workspace)   \10ndvi\project2_mn_ndvi\mn_data_ndvi
RESULTS (Scratch Workspace)   \10ndvi\project2_mn_ndvi\mn_results_ndvi

227

MAKING
SPATIAL
DECISIONS
USING GIS
AND REMOTE
SENSING

3

NORMALIZED
DIFFERENCE
VEGETATION
INDEX

PROJECT 3
On your own

Scenario/problem

You have worked through a project assessing NDVI in Texas and repeated the NDVI analysis for the Minneapolis, Minnesota, region. For this project, you will reinforce your skills by researching and analyzing a similar scenario entirely on your own. First, you must identify your study area and acquire the data for your analysis. You may want to do a local area. Refer to appendix A for directions on how to download your satellite imagery.

Research

Research the problem and answer the following questions:

1. What is the area of study?
2. What is the problem you are going to study?
3. What data is available?

Obtain the data

Do you have access to baseline data? Data and Maps for ArcGIS at http://www.esri.com/data/data-maps provides many of the layers of data that are needed for project work. Be sure to pay particular attention to the source of data and get the latest version. You can obtain data from the following sources:

- USGS Globalization Viewer at http://glovis.usgs.gov: Access to multiple sets of EROS satellite and aerial imagery.

- Census 2000 TIGER/Line Data at http://www.esri.com/tiger: Access to Census 2000 line data.
- Geospatial One-Stop at http://GEO.DATA.GOV: Web-based geospatial resources.
- The National Atlas at http://www.nationalatlas.gov: A range of products and geographic information about the United States.
- The National Map at http://nationalmap.gov/viewer.html: Data includes elevation, land cover, and topographic maps.

Workflow

After researching the problem and obtaining the data, you should do the following:

1. Write a brief scenario.

2. State the problem.

3. Define the deliverables.

4. Examine the metadata.

5. Set the directory structure, start your process summary, and document the map.

6. Decide what you need for the data frame coordinate system and the environments.
   a. What is the best projection for your work?
   b. Do you need to set a cell size or mask?

7. Start your analysis.

8. Prepare your presentation and deliverables.

9. Always remember to document your work in a process summary.

MAKING
SPATIAL
DECISIONS
USING GIS
AND REMOTE
SENSING

3

NORMALIZED
DIFFERENCE
VEGETATION
INDEX

APPENDIX A
DOWNLOADING LANDSAT IMAGERY

MAKING
SPATIAL
DECISIONS
USING GIS
AND REMOTE
SENSING

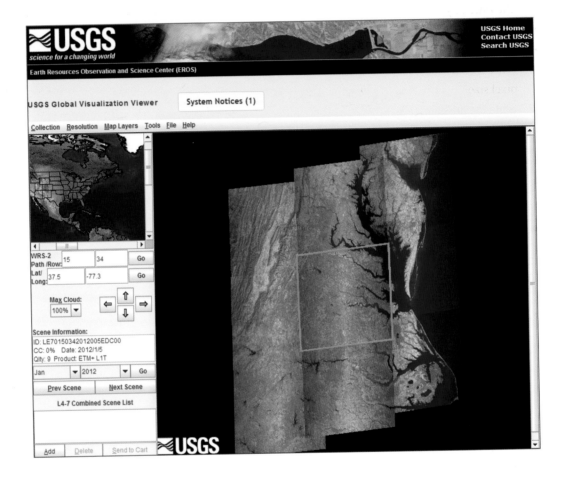

The following instructions are for downloading Landsat imagery from the USGS Global Visualization Viewer (GloVis) website at http://glovis.usgs.gov. Because of the fluid nature of websites, general instructions are given.

When downloading Landsat imagery, the following topics should be considered:

- Thematic Mapper data provides six spectral bands and one thermal band.
- Landsat 5 and Landsat 7 collect TM data.
- Landsat 7 was launched April 15, 1999, but it partially failed in May 2003, causing streaks in the Landsat scenes collected. Therefore, Landsat 5 provides the only continuous source of data.
- The path and row of the Landsat scene needs to be identified.
- The cloud cover needs to be assessed.
- The time of year (leaf on/leaf off) needs to be determined.
- Identify which Landsat collection is needed.
- Level 1 product: Level 1 Landsat is output in a GeoTIFF format, using the universal transverse Mercator (UTM) map projection with a north-up image orientation and a 30-meter pixel size.

MAKING
SPATIAL
DECISIONS
USING GIS
AND REMOTE
SENSING

APPENDIX A

APPENDIX B
REFERENCES

MAKING
SPATIAL
DECISIONS
USING GIS
AND REMOTE
SENSING

Anderson, J. R., E. E. Hardy, J. T. Roach, and R.E. Whitmer. 1976. "A Land Use and Land Cover Classification System for Use with Remote Sensor Data." Geological Survey Professional Paper 964. Washington, DC: US Government Printing Office. http://landcover.usgs.gov/pdf/anderson. pdf.

Boesch, D. F. 2006. "Scientific Requirements for Ecosystem-Based Management in the Restoration of Chesapeake Bay and Coastal Louisiana." *Ecological Engineering* 26 (1): 6–26.

Cresti, R., J. Srivastava, and S. Jung. 2003. *Agriculture Non-Point Source Pollution Control: Good Management Practices—the Chesapeake Bay Experience.* Washington, DC: International Bank for Reconstruction and Development/The World Bank.

Goetz, S. J., C. A. Jantz, S. D. Prince, A. J. Smith, D. Varlyguin, and R. K. Wright. 2004. "Integrated Analysis of Ecosystem Interactions with Land Use Change: Chesapeake Bay Watershed." In *Ecosystems and Land Use Change*, edited by R. S. DeFries, G. P. Asner, and R. A. Houghton. Washington, DC: American Geophysical Union.

Kachhwaha, T. S. 1983. "Spectral Signatures Obtained from Landsat Digital Data for Forest Vegetation and Landuse Mapping in India." *Photogrammetric Engineering and Remote Sensing* 49 (5): 685–89.

Kemp, W. M., W. R. Boynton, J. E. Adolf, D. F. Boesch, W. C. Boicourt, G. Brush, J. C. Cornwell, T. R. Fisher, P. M. Gilbert, J. D. Hagy, L. W. Harding, E. D. Houde, D. G. Kimmel, W. D. Miller, R. I. E. Newell, M. R. Roman, E. M. Smith, and J. C. Stevenson. 2005. "Eutrophication of Chesapeake Bay: Historical Trends and Ecological Interactions." *Marine Ecology Progress Series* 303: 1–29.

Kogan, F. 1997. "Global Drought Watch from Space." *Bulletin of the American Meteorological Society* 78 (4): 621–36.

Xian, G., M. Crane, and C. McMahon. 2008. "Quantifying Multi-temporal Urban Development Characteristics in Las Vegas from Landsat and ASTER Data." *Photogrammetric Engineering and Remote Sensing* 74 (4): 473–81.

Yang, L., C. Huang, C. G. Homer, B. K. Wylie, and M. J. Coan. 2003. "An Approach for Mapping Large-Area Impervious Surfaces: Synergistic Use of Landsat-7 ETM+ and High Spatial Resolution Imagery." *Canadian Journal of Remote Sensing* 29 (2): 230–40.

MAKING

SPATIAL

DECISIONS

USING GIS

AND REMOTE

SENSING

APPENDIX B

APPENDIX C
IMAGE AND DATA CREDITS

MAKING
SPATIAL
DECISIONS
USING GIS
AND REMOTE
SENSING

All images created by Keranen and Kolvoord.

Basemap from ArcGIS Online (figure 3a-1), world imagery courtesy of Esri, i-cubed, USDA, USGS, AEX, GeoEye, Getmapping, Aerogrid, IGN, IGP, and the GIS user community.

Basemap from ArcGIS Online (figure 3a-2), world imagery courtesy of Esri, i-cubed, USDA, USGS, AEX, GeoEye, Getmapping, Aerogrid, IGN, IGP, and the GIS user community.

Basemap from ArcGIS Online (figure 3a-3), world imagery courtesy of Esri, i-cubed, USDA, USGS, AEX, GeoEye, Getmapping, Aerogrid, IGN, IGP, and the GIS user community.

Land-cover data from ArcGIS Online (figure 4a-1), image courtesy of US Geological Survey.

Image (figure 4a-2) from ArcGIS Help 10.1 http://resources.arcgis.com/en/help/main/10.1/index.html#/How_features_are_represented_in_a_raster/009t00000006000000/, courtesy of Esri.

Basemap from ArcGIS Online (figure 4a-3), world imagery courtesy of Esri, i-cubed, USDA, USGS, AEX, GeoEye, Getmapping, Aerogrid, IGN, IGP, and the GIS user community.

Basemap from ArcGIS Online (figure 4b-1), world imagery courtesy of Esri, i-cubed, USDA, USGS, AEX, GeoEye, Getmapping, Aerogrid, IGN, IGP, and the GIS user community.

Basemap from ArcGIS Online (figure 5a-3), terrain imagery courtesy of USGS, Esri, TANA, and AND.

Basemap from ArcGIS Online (figure 7a-2), world imagery courtesy of Esri, i-cubed, USDA, USGS, AEX, GeoEye, Getmapping, Aerogrid, IGN, IGP, and the GIS user community.

Basemap from ArcGIS Online (figure 7a-3), world imagery courtesy of Esri, i-cubed, USDA, USGS, AEX, GeoEye, Getmapping, Aerogrid, IGN, IGP, and the GIS user community.

Basemap from ArcGIS Online (figure 7a-4), world imagery courtesy of Esri, i-cubed, USDA, USGS, AEX, GeoEye, Getmapping, Aerogrid, IGN, IGP, and the GIS user community.

Basemap from ArcGIS Online (figure 9a-2), world imagery courtesy of Esri, i-cubed, USDA, USGS, AEX, GeoEye, Getmapping, Aerogrid, IGN, IGP, and the GIS user community.

Basemap from ArcGIS Online (figure 9b-3), world imagery courtesy of Esri, i-cubed, USDA, USGS, AEX, GeoEye, Getmapping, Aerogrid, IGN, IGP, and the GIS user community.

Basemap from ArcGIS Online (figure 9b-4), world imagery courtesy of Esri, i-cubed, USDA, USGS, AEX, GeoEye, Getmapping, Aerogrid, IGN, IGP, and the GIS user community.

Landsat image (appendix A), courtesy of the US Geological Survey and Keranen and Kolvoord.

Data credits

Module 1, project 1

\EsriPress\MSDRemSen\Data\01enhance\project1_bay_enhance\bay_data_enhance\landsat_may2006\L5015033_03320060504_MTL, image courtesy of the US Geological Survey.

\EsriPress\MSDRemSen\Data\01enhance\project1_bay_enhance\bay_data_enhance\bay.gdb\bayfeatures\AOI, Keranen and Kolvoord.

\EsriPress\MSDRemSen\Data\01enhance\project1_bay_enhance\bay_data_enhance\bay.gdb\bayfeatures\dd_studyarea, Keranen and Kolvoord.

\EsriPress\MSDRemSen\Data\01enhance\project1_bay_enhance\bay_data_enhance\bay.gdb\bayfeatures\highways, from Data and Maps for ArcGIS courtesy of Esri.

\EsriPress\MSDRemSen\Data\01enhance\project1_bay_enhance\bay_data_enhance\bay.gdb\bayfeatures\rivers, from Data and Maps for ArcGIS courtesy of the US Geological Survey and Esri.

\EsriPress\MSDRemSen\Data\01enhance\project1_bay_enhance\bay_data_enhance\bay.gdb\bayfeatures\states, from Data and Maps for ArcGIS courtesy of ArcUSA, US Census, Esri.

\EsriPress\MSDRemSen\Data\01enhance\project1_bay_enhance\bay_data_enhance\bay.gdb\bayfeatures\watershed, image courtesy of the US Geological Survey.

\EsriPress\MSDRemSen\Data\01enhance\project1_bay_enhance\bay_data_enhance\bay.gdb\bayfeatures\clip_band4, image courtesy of the US Geological Survey and modified by Keranen and Kolvoord.

MAKING
SPATIAL
DECISIONS
USING GIS
AND REMOTE
SENSING

APPENDIX C

\EsriPress\MSDRemSen\Data\01enhance\project1_bay_enhance\bay_data_enhance\bay.
gdb\bayfeatures\highways.lyr, from Data and Maps for ArcGIS http://www.esri.com/data/
data-maps/data-and-maps-dvd courtesy of Esri.

\EsriPress\MSDRemSen\Data\01enhance\project1_bay_enhance\bay_results_enhance\bay_
results.gdb, Keranen and Kolvoord.

MAKING
SPATIAL
DECISIONS
USING GIS
AND REMOTE
SENSING

APPENDIX C

Module 1, project 2

\EsriPress\MSDRemSen\Data\01enhance\project2_vegas_enhance\vegas_data_enhance\
landsat_may_2006\L5039035_03520060512_MTL, image courtesy of the US Geological Survey.

\EsriPress\MSDRemSen\Data\01enhance\project2_vegas_enhance\vegas_data_enhance\
vegas.gdb\AOI, Keranen and Kolvoord.

\EsriPress\MSDRemSen\Data\01enhance\project2_vegas_enhance\vegas_data_enhance\
vegas.gdb\dd_studyarea, Keranen and Kolvoord.

\EsriPress\MSDRemSen\Data\01enhance\project2_vegas_enhance\vegas_data_enhance\
vegas.gdb\dtl_riv, courtesy of the US Geological Survey and Esri.

\EsriPress\MSDRemSen\Data\01enhance\project2_vegas_enhance\vegas_data_enhance\
vegas.gdb\dtl_wat, image courtesy of the US Geological Survey and Esri.

\EsriPress\MSDRemSen\Data\01enhance\project2_vegas_enhance\vegas_data_enhance\
vegas.gdb\index_150, Keranen and Kolvoord.

\EsriPress\MSDRemSen\Data\01enhance\project2_vegas_enhance\vegas_data_enhance\
vegas.gdb\mjr_hwys, from Data and Maps for ArcGIS courtesy of Esri.

\EsriPress\MSDRemSen\Data\01enhance\project2_vegas_enhance\vegas_data_enhance\
vegas.gdb\states, from Data and Maps for ArcGIS courtesy of ArcUSA, US Census, ESRI.

\EsriPress\MSDRemSen\Data\01enhance\project2_vegas_enhance\vegas_data_enhance\
vegas.gdb\urban, courtesy of US Census Bureau.

\EsriPress\MSDRemSen\Data\01enhance\project2_vegas_enhance\vegas_data_enhance\
vegas.gdb\clip_band4, image courtesy of the US Geological Survey and modified by Keranen
and Kolvoord.

\EsriPress\MSDRemSen\Data\01enhance\project2_vegas_enhance\vegas_data_enhance\
vegas.gdb\highways.lyr, from Data and Maps for ArcGIS courtesy of Esri.

\EsriPress\MSDRemSen\Data\01enhance\project2_vegas_enhance\vegas_results_enhance\
vegas_results.gdb, Keranen and Kolvoord.

MAKING
SPATIAL
DECISIONS
USING GIS
AND REMOTE
SENSING

Module 2, project 1

\EsriPress\MSDRemSen\Data\02composite\project1_bay_comp\bay_data_comp\landsat_
may_2006\L5015033_03320060504_MTL, image courtesy of the US Geological Survey.

\EsriPress\MSDRemSen\Data\02composite\project1_bay_comp\bay_data_comp\bay.gdb\
bayfeatures\AOI, Keranen and Kolvoord.

\EsriPress\MSDRemSen\Data\02composite\project1_bay_comp\bay_data_comp\bay.gdb\
bayfeatures\dd_studyarea, Keranen and Kolvoord.

\EsriPress\MSDRemSen\Data\02composite\project1_bay_comp\bay_data_comp\bay.gdb\
bayfeatures\highways, from Data and Maps for ArcGIS courtesy of Esri.

\EsriPress\MSDRemSen\Data\02composite\project1_bay_comp\bay_data_comp\bay.gdb\
bayfeatures\rivers, from Data and Maps for ArcGIS courtesy of the US Geological Survey and
Esri.

\EsriPress\MSDRemSen\Data\02composite\project1_bay_comp\bay_data_comp\bay.gdb\
bayfeatures\states, from Data and Maps for ArcGIS courtesy of ArcUSA, US Census, Esri.

\EsriPress\MSDRemSen\Data\02composite\project1_bay_comp\bay_data_comp\bay.gdb\
bayfeatures\watershed, image courtesy of the US Geological Survey.

\EsriPress\MSDRemSen\Data\02composite\project1_bay_comp\bay_data_comp\bay.gdb\clip_
band4, image courtesy of the US Geological Survey and modified by Keranen and Kolvoord.

\EsriPress\MSDRemSen\Data\02composite\project1_bay_comp\bay_data_comp\bay.gdb\
highways.lyr, from Data and Maps for ArcGIS courtesy of Esri.

\EsriPress\MSDRemSen\Data\02composite\project1_bay_comp\bay_results_comp\bay_
results.gdb, Keranen and Kolvoord.

\EsriPress\MSDRemSen\Data\02composite\project1_bay_comp\bay_data_comp\landsat_may_2006\L5015033_03320060504_MTL, image courtesy of the US Geological Survey.

\EsriPress\MSDRemSen\Data\02composite\project1_bay_comp\bay_data_comp\bay.gdb\bayfeatures\AOI, Keranen and Kolvoord.

MAKING
SPATIAL
DECISIONS
USING GIS
AND REMOTE
SENSING

APPENDIX C

\EsriPress\MSDRemSen\Data\02composite\project1_bay_comp\bay_data_comp\bay.gdb\bayfeatures\dd_studyarea, Keranen and Kolvoord.

\EsriPress\MSDRemSen\Data\02composite\project1_bay_comp\bay_data_comp\bay.gdb\bayfeatures\highways, from Data and Maps for ArcGIS courtesy of Esri.

\EsriPress\MSDRemSen\Data\02composite\project1_bay_comp\bay_data_comp\bay.gdb\bayfeatures\rivers, from Data and Maps for ArcGIS courtesy of US Geological Survey and Esri.

\EsriPress\MSDRemSen\Data\02composite\project1_bay_comp\bay_data_comp\bay.gdb\bayfeatures\states, from Data and Maps for ArcGIS courtesy of ArcUSA, US Census, Esri.

\EsriPress\MSDRemSen\Data\02composite\project1_bay_comp\bay_data_comp\bay.gdb\bayfeatures\watershed, image courtesy of the US Geological Survey.

\EsriPress\MSDRemSen\Data\02composite\project1_bay_comp\bay_data_comp\bay.gdb\clip_band4, image courtesy of the US Geological Survey and modified by Keranen and Kolvoord.

\EsriPress\MSDRemSen\Data\02composite\project1_bay_comp\bay_data_comp\bay.gdb\highways.lyr, from Data and Maps for ArcGIS courtesy of Esri.

\EsriPress\MSDRemSen\Data\02composite\project1_bay_comp\bay_results_comp\bay_results.gdb, Keranen and Kolvoord.

Module 2, project 2

\EsriPress\MSDRemSen\Data\02composite\project2_vegas_comp\vegas_data_comp\vegas.gdb\vegasfeatures\AOI, Keranen and Kolvoord.

\EsriPress\MSDRemSen\Data\02composite\project2_vegas_comp\vegas_data_comp\vegas.gdb\vegasfeatures\dd_studyarea, Keranen and Kolvoord.

\EsriPress\MSDRemSen\Data\02composite\project2_vegas_comp\vegas_data_comp\vegas.gdb\vegasfeatures\dtl_riv, image courtesy of the US Geological Survey and Esri.

\EsriPress\MSDRemSen\Data\02composite\project2_vegas_comp\vegas_data_comp\vegas.gdb\vegasfeatures\dtl_wat, image courtesy of the US Geological Survey and Esri.

\EsriPress\MSDRemSen\Data\02composite\project2_vegas_comp\vegas_data_comp\vegas.gdb\vegasfeatures\mjr_hwys, from Data and Maps for ArcGIS courtesy of Esri.

\EsriPress\MSDRemSen\Data\02composite\project2_vegas_comp\vegas_data_comp\vegas.gdb\vegasfeatures\states, from Data and Maps for ArcGIS courtesy of ArcUSA, US Census Bureau, Esri.

\EsriPress\MSDRemSen\Data\02composite\project2_vegas_comp\vegas_data_comp\vegas.gdb\vegasfeatures\urban, courtesy of US Census Bureau.

\EsriPress\MSDRemSen\Data\02composite\project2_vegas_comp\vegas_data_comp\vegas.gdb\clip_band4, image courtesy of the US Geological Survey and modified by Keranen and Kolvoord.

\EsriPress\MSDRemSen\Data\02composite\project2_vegas_comp\vegas_data_comp\vegas.gdb\highways.lyr, from Data and Maps for ArcGIS courtesy of Esri.

\EsriPress\MSDRemSen\Data\02composite\project2_vegas_comp\vegas_results_comp\vegas_results.gdb, Keranen and Kolvoord.

MAKING
SPATIAL
DECISIONS
USING GIS
AND REMOTE
SENSING

APPENDIX C

Module 3, project 1

\EsriPress\MSDRemSen\Data\03signatures\project1_bay_ss\bay_data_ss\landsat_may_2006\L5015033_03320060504_MTL, image courtesy of the US Geological Survey.

\EsriPress\MSDRemSen\Data\03signatures\project1_bay_ss\bay_data_ss\bay.gdb\bayfeatures\AOI, Keranen and Kolvoord.

\EsriPress\MSDRemSen\Data\03signatures\project1_bay_ss\bay_data_ss\bay.gdb\bayfeatures\dd_studyarea, Keranen and Kolvoord.

\EsriPress\MSDRemSen\Data\03signatures\project1_bay_ss\bay_data_ss\bay.gdb\bayfeatures\highways, from Data and Maps for ArcGIS courtesy of Esri.

\EsriPress\MSDRemSen\Data\03signatures\project1_bay_ss\bay_data_ss\bay.gdb\bayfeatures\rivers, from Data and Maps for ArcGIS courtesy of the US Geological Survey and Esri.

\EsriPress\MSDRemSen\Data\03signatures\project1_bay_ss\bay_data_ss\bay.gdb\bayfeatures\states, from Data and Maps for ArcGIS courtesy of ArcUSA, US Census, Esri.

MAKING
SPATIAL
DECISIONS
USING GIS
AND REMOTE
SENSING

APPENDIX C

\EsriPress\MSDRemSen\Data\03signatures\project1_bay_ss\bay_data_ss\bay.gdb\bayfeatures\ watershed, image courtesy of the US Geological Survey.

\EsriPress\MSDRemSen\Data\03signatures\project1_bay_ss\bay_data_ss\bay.gdb\bayfeatures\ clear_water, Keranen and Kolvoord.

\EsriPress\MSDRemSen\Data\03signatures\project1_bay_ss\bay_data_ss\bay.gdb\ss_points\ con_trees, Keranen and Kolvoord.

\EsriPress\MSDRemSen\Data\03signatures\project1_bay_ss\bay_data_ss\bay.gdb\ss_points\ concrete, Keranen and Kolvoord.

\EsriPress\MSDRemSen\Data\03signatures\project1_bay_ss\bay_data_ss\bay.gdb\ss_points\ dec_trees, Keranen and Kolvoord.

\EsriPress\MSDRemSen\Data\03signatures\project1_bay_ss\bay_data_ss\bay.gdb\ss_points\ grass, Keranen and Kolvoord.

\EsriPress\MSDRemSen\Data\03signatures\project1_bay_ss\bay_data_ss\bay.gdb\ss_points\ shrub_scrub, Keranen and Kolvoord.

\EsriPress\MSDRemSen\Data\03signatures\project1_bay_ss\bay_data_ss\bay.gdb\ss_points\ turbid_water, Keranen and Kolvoord.

\EsriPress\MSDRemSen\Data\03signatures\project1_bay_ss\bay_data_ss\bay.gdb\ss_points\ water, Keranen and Kolvoord.

\EsriPress\MSDRemSen\Data\03signatures\project1_bay_ss\bay_data_ss\bay.gdb\ clip_band4, image courtesy of the US Geological Survey and modified by Keranen and Kolvoord.

\EsriPress\MSDRemSen\Data\03signatures\project1_bay_ss\bay_data_ss\bay.gdb\highways. lyr, from Data and Maps for ArcGIS courtesy of Esri.

\EsriPress\MSDRemSen\Data\03signatures\project1_bay_ss\bay_results_ss\bay_results.gdb, Keranen and Kolvoord.

\EsriPress\MSDRemSen\Data\03signatures\project1_bay_ss\Documents\table1.xlsx, Keranen and Kolvoord.

\EsriPress\MSDRemSen\Data\03signatures\project1_bay_ss\Documents\table2.xlsx, Keranen and Kolvoord.

\EsriPress\MSDRemSen\Data\03signatures\project1_bay_ss\Documents\table3.xlsx, Keranen and Kolvoord.

MAKING
SPATIAL
DECISIONS
USING GIS
AND REMOTE
SENSING

Module 3, project 2

\EsriPress\MSDRemSen\Data\03signatures\project2_vegas_ss\documents\table1.xlsx, Keranen and Kolvoord.

\EsriPress\MSDRemSen\Data\03signatures\project2_vegas_ss\documents\table2.xlsx, Keranen and Kolvoord.

\EsriPress\MSDRemSen\Data\03signatures\project2_vegas_ss\documents\table3.xlsx, Keranen and Kolvoord.

\EsriPress\MSDRemSen\Data\03signatures\project2_vegas_ss\vegas_data_ss\landsat_may_2006\L5039035_03520060512_MTL, image courtesy of the US Geological Survey.

\EsriPress\MSDRemSen\Data\clear_water, Keranen and Kolvoord.

\EsriPress\MSDRemSen\Data\03signatures\project2_vegas_ss\vegas_data_ss\Vegas.gdb\ss_points\clear_water, Keranen and Kolvoord.

\EsriPress\MSDRemSen\Data\03signatures\project2_vegas_ss\vegas_data_ss\Vegas.gdb\ss_points\concrete, Keranen and Kolvoord.

\EsriPress\MSDRemSen\Data\03signatures\project2_vegas_ss\vegas_data_ss\Vegas.gdb\ss_points\evergreen, Keranen and Kolvoord.

\EsriPress\MSDRemSen\Data\03signatures\project2_vegas_ss\vegas_data_ss\Vegas.gdb\ss_points\grass, Keranen and Kolvoord.

\EsriPress\MSDRemSen\Data\03signatures\project2_vegas_ss\vegas_data_ss\Vegas.gdb\ss_points\turbid_water, Keranen and Kolvoord.

\EsriPress\MSDRemSen\Data\03signatures\project2_vegas_ss\vegas_data_ss\Vegas.gdb\ss_points\water, Keranen and Kolvoord.

MAKING
SPATIAL
DECISIONS
USING GIS
AND REMOTE
SENSING

APPENDIX C

\EsriPress\MSDRemSen\Data\03signatures\project2_vegas_ss\vegas_data_ss\Vegas.gdb\vegas\AOI, Keranen and Kolvoord.

\EsriPress\MSDRemSen\Data\03signatures\project2_vegas_ss\vegas_data_ss\Vegas.gdb\vegas\dd_studyarea, Keranen and Kolvoord.

\EsriPress\MSDRemSen\Data\03signatures\project2_vegas_ss\vegas_data_ss\Vegas.gdb\vegas\dtl_river, courtesy of the US Geological Survey and Esri.

\EsriPress\MSDRemSen\Data\03signatures\project2_vegas_ss\vegas_data_ss\Vegas.gdb\vegas\dtl_wat, image courtesy of the US Geological Survey and Esri.

\EsriPress\MSDRemSen\Data\03signatures\project2_vegas_ss\vegas_data_ss\Vegas.gdb\vegas\index_50, Keranen and Kolvoord.

\EsriPress\MSDRemSen\Data\03signatures\project2_vegas_ss\vegas_data_ss\Vegas.gdb\vegas\mjr_hwys, from Data and Maps for ArcGIS courtesy of Esri.

\EsriPress\MSDRemSen\Data\03signatures\project2_vegas_ss\vegas_data_ss\Vegas.gdb\vegas\states, from Data and Maps for ArcGIS courtesy of ArcUSA, US Census, ESRI.

\EsriPress\MSDRemSen\Data\03signatures\project2_vegas_ss\vegas_data_ss\Vegas.gdb\vegas\urban, courtesy of US Census Bureau.

\EsriPress\MSDRemSen\Data\03signatures\project2_vegas_ss\vegas_data_ss\highways.lyr, from Data and Maps for ArcGIS courtesy of Esri.

\EsriPress\MSDRemSen\Data\03signatures\project2_vegas_ss\vegas_results_ss\vegas_results_ss.gdb, Keranen and Kolvoord.

Module 4, project 1

\EsriPress\MSDRemSen\Data\04landcover\project1_loudoun\loudoun_data\bay.gdb\bay features\AOI, Keranen and Kolvoord.

\EsriPress\MSDRemSen\Data\04landcover\project1_loudoun\loudoun_data\bay.gdb\bay features\BD_studyarea, Keranen and Kolvoord.

\EsriPress\MSDRemSen\Data\04landcover\project1_loudoun\loudoun_data\bay.gdb\bay features\Beaverdam, Keranen and Kolvoord.

\EsriPress\MSDRemSen\Data\04landcover\project1_loudoun\loudoun_data\bay.gdb\bay features\dd_studyarea, Keranen and Kolvoord.

\EsriPress\MSDRemSen\Data\04landcover\project1_loudoun\loudoun_data\bay.gdb\bay features\digitize, Keranen and Kolvoord.

\EsriPress\MSDRemSen\Data\04landcover\project1_loudoun\loudoun_data\bay.gdb\bay features\highways, from Data and Maps for ArcGIS courtesy of Esri.

\EsriPress\MSDRemSen\Data\04landcover\project1_loudoun\loudoun_data\bay.gdb\bay features\Loudoun, Keranen and Kolvoord.

\EsriPress\MSDRemSen\Data\04landcover\project1_loudoun\loudoun_data\bay.gdb\bay features\rivers, from Data and Maps for ArcGIS courtesy of the US Geological Survey and Esri.

\EsriPress\MSDRemSen\Data\04landcover\project1_loudoun\loudoun_data\bay.gdb\bay features\states, from Data and Maps for ArcGIS courtesy of ArcUSA, US Census, Esri.

\EsriPress\MSDRemSen\Data\04landcover\project1_loudoun\loudoun_data\bay.gdb\bay features\watershed, image courtesy of the US Geological Survey.

\EsriPress\MSDRemSen\Data\04landcover\project1_loudoun\loudoun_data\bay.gdb\LC_rec, Keranen and Kolvoord.

\EsriPress\MSDRemSen\Data\04landcover\project1_loudoun\loudoun_data\bay.gdb\highways.lyr, from Data and Maps for ArcGIS courtesy of Esri.

\EsriPress\MSDRemSen\Data\04landcover\project1_loudoun\loudoun_results\loudoun_results.gdb, Keranen and Kolvoord.

MAKING
SPATIAL
DECISIONS
USING GIS
AND REMOTE
SENSING

APPENDIX C

Module 4, project 2

\EsriPress\MSDRemSen\Data\04landcover\project2_mohave\mohave_data\landsat_may_2006\L5039035_03520060512_MTL, image courtesy of the US Geological Survey.

\EsriPress\MSDRemSen\Data\04landcover\project2_mohave\mohave_data\mohave.gdb\Vegas\Digitize, Keranen and Kolvoord.

\EsriPress\MSDRemSen\Data\04landcover\project2_mohave\mohave_data\mohave.gdb\Vegas\dtl_river, courtesy of the US Geological Survey and Esri.

MAKING
SPATIAL
DECISIONS
USING GIS
AND REMOTE
SENSING

APPENDIX C

\EsriPress\MSDRemSen\Data\04landcover\project2_mohave\mohave_data\mohave.gdb\ Vegas\Lake_Mohave, Keranen and Kolvoord.

\EsriPress\MSDRemSen\Data\04landcover\project2_mohave\mohave_data\mohave.gdb\ Vegas\mjr_hwys, from Data and Maps for ArcGIS courtesy of Esri.

\EsriPress\MSDRemSen\Data\04landcover\project2_mohave\mohave_data\mohave.gdb\ Vegas\states, from Data and Maps for ArcGIS courtesy of ArcUSA, US Census, Esri.

\EsriPress\MSDRemSen\Data\04landcover\project2_mohave\mohave_data\mohave.gdb\ Vegas\study_area, Keranen and Kolvoord.

\EsriPress\MSDRemSen\Data\04landcover\project2_mohave\mohave_data\mohave.gdb\ Vegas\urban, courtesy of US Census Bureau.

\EsriPress\MSDRemSen\Data\04landcover\project2_mohave\mohave_data\mohave.gdb\LC_ rec, Keranen and Kolvoord.

\EsriPress\MSDRemSen\Data\04landcover\project2_mohave\mohave_data\mohave.gdb\ highways.lyr, from Data and Maps for ArcGIS courtesy of Esri.

\EsriPress\MSDRemSen\Data\04landcover\project2_mohave\mohave_results\mohave_results. gdb, Keranen and Kolvoord.

Module 5, project 1

\EsriPress\MSDRemSen\Data\05unsupervised\project1_bay_unsup\bay_data_unsup\landsat_ may_2006\L5015033_03320060504_MTL, image courtesy of the US Geological Survey.

\EsriPress\MSDRemSen\Data\05unsupervised\project1_bay_unsup\bay_data_unsup\bay.gdb\ bayfeatures\AOI, Keranen and Kolvoord.

\EsriPress\MSDRemSen\Data\05unsupervised\project1_bay_unsup\bay_data_unsup\bay.gdb\ bayfeatures\dd_studyarea, Keranen and Kolvoord.

\EsriPress\MSDRemSen\Data\05unsupervised\project1_bay_unsup\bay_data_unsup\bay.gdb\ bayfeatures\highways, from Data and Maps for ArcGIS courtesy of Esri.

\EsriPress\MSDRemSen\Data\05unsupervised\project1_bay_unsup\bay_data_unsup\bay.gdb\ bayfeatures\rivers, from Data and Maps for ArcGIS courtesy of the US Geological Survey and Esri.

\EsriPress\MSDRemSen\Data\05unsupervised\project1_bay_unsup\bay_data_unsup\bay.gdb\ bayfeatures\sel_sheds, image courtesy of the US Geological Survey.

\EsriPress\MSDRemSen\Data\05unsupervised\project1_bay_unsup\bay_data_unsup\bay.gdb\ bayfeatures\shed_highways, from Data and Maps for ArcGIS courtesy of Esri.

MAKING
SPATIAL
DECISIONS
USING GIS
AND REMOTE
SENSING

\EsriPress\MSDRemSen\Data\05unsupervised\project1_bay_unsup\bay_data_unsup\bay.gdb\ bayfeatures\shed_rivers, from Data and Maps for ArcGIS courtesy of the US Geological Survey and Esri.

\EsriPress\MSDRemSen\Data\05unsupervised\project1_bay_unsup\bay_data_unsup\bay.gdb\ bayfeatures\states, from Data and Maps for ArcGIS courtesy of ArcUSA, US Census, Esri.

\EsriPress\MSDRemSen\Data\05unsupervised\project1_bay_unsup\bay_data_unsup\bay.gdb\ bayfeatures\watershed, image courtesy of the US Geological Survey.

\EsriPress\MSDRemSen\Data\05unsupervised\project1_bay_unsup\bay_data_unsup\bay. gdb\clip_band4, image courtesy of the US Geological Survey and modified by Keranen and Kolvoord.

\EsriPress\MSDRemSen\Data\05unsupervised\project1_bay_unsup\bay_data_unsup\bay.gdb\ highways.lyr, from Data and Maps for ArcGIS courtesy of Esri.

\EsriPress\MSDRemSen\Data\05unsupervised\project1_bay_unsup\bay_results_unsup\ unsupbay_results.gdb, Keranen and Kolvoord.

Module 5, project 2

\EsriPress\MSDRemSen\Data\05unsupervised\project2_vegas_unsup\vegas_data_unsup\ landsat_may_2006\L5039035_03520060512_MTL, image courtesy of the US Geological Survey.

\EsriPress\MSDRemSen\Data\05unsupervised\project2_vegas_unsup\vegas_data_unsup\ vegas.gdb\Vegas\dtl.river, Keranen and Kolvoord.

\EsriPress\MSDRemSen\Data\05unsupervised\project2_vegas_unsup\vegas_data_unsup\ vegas.gdb\Vegas\dtl_wat, image courtesy of the US Geological Survey and Esri.

\EsriPress\MSDRemSen\Data\05unsupervised\project2_vegas_unsup\vegas_data_unsup\ vegas.gdb\Vegas\mjr_hwys, from Data and Maps for ArcGIS courtesy of Esri.

MAKING
SPATIAL
DECISIONS
USING GIS
AND REMOTE
SENSING

APPENDIX C

\EsriPress\MSDRemSen\Data\05unsupervised\project2_vegas_unsup\vegas_data_unsup\
vegas.gdb\Vegas\states, from Data and Maps for ArcGIS courtesy of ArcUSA, US Census, Esri.

\EsriPress\MSDRemSen\Data\05unsupervised\project2_vegas_unsup\vegas_data_unsup\
vegas.gdb\Vegas\study_area, Keranen and Kolvoord.

\EsriPress\MSDRemSen\Data\05unsupervised\project2_vegas_unsup\vegas_data_unsup\
vegas.gdb\Vegas\urban, courtesy of US Census Bureau.

\EsriPress\MSDRemSen\Data\05unsupervised\project2_vegas_unsup\vegas_data_unsup\
vegas.gdb\Vegas\watersheds, image courtesy of the US Geological Survey.

\EsriPress\MSDRemSen\Data\05unsupervised\project2_vegas_unsup\vegas_data_unsup\
vegas.gdb\highways.lyr, from Data and Maps for ArcGIS courtesy of Esri.

\EsriPress\MSDRemSen\Data\05unsupervised\project2_vegas_unsup\vegas_results_unsup\
vegas_results.gdb, Keranen and Kolvoord.

Module 6, project 1

\EsriPress\MSDRemSen\Data\06supervised\project1_bay_sup\bay_data_sup\landsat_
may_2006\L5015033_03320060504_MTL, image courtesy of the US Geological Survey.

\EsriPress\MSDRemSen\Data\06supervised\project1_bay_sup\bay_data_sup\bay.gdb\bay
features\AOI, Keranen and Kolvoord.

\EsriPress\MSDRemSen\Data\06supervised\project1_bay_sup\bay_data_sup\bay.gdb\bay
features\dd_studyarea, Keranen and Kolvoord.

\EsriPress\MSDRemSen\Data\06supervised\project1_bay_sup\bay_data_sup\bay.gdb\bay
features\highways, from Data and Maps for ArcGIS courtesy of Esri.

\EsriPress\MSDRemSen\Data\06supervised\project1_bay_sup\bay_data_sup\bay.gdb\bay
features\rivers, from Data and Maps for ArcGIS courtesy of the US Geological Survey and
Esri.

\EsriPress\MSDRemSen\Data\06supervised\project1_bay_sup\bay_data_sup\bay.gdb\bay
features\sel_sheds, image courtesy of the US Geological Survey.

\EsriPress\MSDRemSen\Data\06supervised\project1_bay_sup\bay_data_sup\bay.gdb\bay
features\shed_highways, from Data and Maps for ArcGIS courtesy of Esri.

\EsriPress\MSDRemSen\Data\06supervised\project1_bay_sup\bay_data_sup\bay.gdb\bay features\shed_rivers, from Data and Maps for ArcGIS courtesy of the US Geological Survey and Esri.

\EsriPress\MSDRemSen\Data\06supervised\project1_bay_sup\bay_data_sup\bay.gdb\bay features\states, from Data and Maps for ArcGIS courtesy of ArcUSA, US Census, Esri.

\EsriPress\MSDRemSen\Data\06supervised\project1_bay_sup\bay_data_sup\bay.gdb\bay features\watershed, image courtesy of the US Geological Survey.

\EsriPress\MSDRemSen\Data\06supervised\project1_bay_sup\bay_data_sup\bay.gdb\clip_band4, image courtesy of the US Geological Survey and modified by Keranen and Kolvoord.

\EsriPress\MSDRemSen\Data\06supervised\project1_bay_sup\bay_data_sup\bay.gdb\highways.lyr, from Data and Maps for ArcGIS courtesy of Esri.

EsriPress\MSDRemSen\Data\06supervised\project1_bay_sup\bay_results_sup\bay_results_sup.gdb, Keranen and Kolvoord.

Module 6, project 2

\EsriPress\MSDRemSen\Data\06supervised\project2_vegas_sup\vegas_data_sup\landsat_may_2006\L5039035_03520060512_MTL, image courtesy of the US Geological Survey.

\EsriPress\MSDRemSen\Data\06supervised\project2_vegas_sup\vegas_data_sup\vegas.gdb\Vegas\dtl.river, courtesy of the US Geological Survey and Esri.

\EsriPress\MSDRemSen\Data\06supervised\project2_vegas_sup\vegas_data_sup\vegas.gdb\Vegas\dtl_wat, image courtesy of the US Geological Survey and Esri.

\EsriPress\MSDRemSen\Data\06supervised\project2_vegas_sup\vegas_data_sup\vegas.gdb\Vegas\mjr_hwys, from Data and Maps for ArcGIS courtesy of Esri.

\EsriPress\MSDRemSen\Data\06supervised\project2_vegas_sup\vegas_data_sup\vegas.gdb\Vegas\states, from Data and Maps for ArcGIS courtesy of ArcUSA, US Census, Esri.

\EsriPress\MSDRemSen\Data\06supervised\project2_vegas_sup\vegas_data_sup\vegas.gdb\Vegas\studyarea, Keranen and Kolvoord.

\EsriPress\MSDRemSen\Data\06supervised\project2_vegas_sup\vegas_data_sup\vegas.gdb\Vegas\urban, courtesy of US Census Bureau.

\EsriPress\MSDRemSen\Data\06supervised\project2_vegas_sup\vegas_data_sup\vegas.gdb\
Vegas\watersheds, image courtesy of the US Geological Survey.

\EsriPress\MSDRemSen\Data\06supervised\project2_vegas_sup\vegas_data_sup\vegas.gdb\
highways.lyr, from Data and Maps for ArcGIS courtesy of Esri.

MAKING
SPATIAL
DECISIONS
USING GIS
AND REMOTE
SENSING

\EsriPress\MSDRemSen\Data\06supervised\project2_vegas_sup\vegas_data_sup\vegas.gdb\
landcover.lyr, Keranen and Kolvoord.

\EsriPress\MSDRemSen\Data\06supervised\project2_vegas_sup\vegas_results_sup\vegas_
results.gdb, Keranen and Kolvoord.

Module 7, project 1

\EsriPress\MSDRemSen\Data\07accuracy\project1_bay_gt\bay_data_gt\landsat_may_2006\
L5015033_03320060504_MTL, image courtesy of the US Geological Survey.

\EsriPress\MSDRemSen\Data\07accuracy\project1_bay_gt\bay_data_gt\bay.gdb\bayfeatures\
AOI, Keranen and Kolvoord.

\EsriPress\MSDRemSen\Data\07accuracy\project1_bay_gt\bay_data_gt\bay.gdb\bayfeatures\
dd_studyarea, Keranen and Kolvoord.

\EsriPress\MSDRemSen\Data\07accuracy\project1_bay_gt\bay_data_gt\bay.gdb\bayfeatures\
highways, from Data and Maps for ArcGIS courtesy of Esri.

\EsriPress\MSDRemSen\Data\07accuracy\project1_bay_gt\bay_data_gt\bay.gdb\bayfeatures\
rivers, from Data and Maps for ArcGIS courtesy of the US Geological Survey and Esri.

\EsriPress\MSDRemSen\Data\07accuracy\project1_bay_gt\bay_data_gt\bay.gdb\bayfeatures\
sel_sheds, image courtesy of the US Geological Survey.

\EsriPress\MSDRemSen\Data\07accuracy\project1_bay_gt\bay_data_gt\bay.gdb\bayfeatures\
shed_highways, from Data and Maps for ArcGIS courtesy of Esri.

\EsriPress\MSDRemSen\Data\07accuracy\project1_bay_gt\bay_data_gt\bay.gdb\bayfeatures\
shed_rivers, from Data and Maps for ArcGIS courtesy of the US Geological Survey and Esri.

\EsriPress\MSDRemSen\Data\07accuracy\project1_bay_gt\bay_data_gt\bay.gdb\bayfeatures\
states, from Data and Maps for ArcGIS courtesy of ArcUSA, US Census, Esri.

\EsriPress\MSDRemSen\Data\07accuracy\project1_bay_gt\bay_data_gt\bay.gdb\bayfeatures\ watershed, image courtesy of the US Geological Survey.

\EsriPress\MSDRemSen\Data\07accuracy\project1_bay_gt\bay_data_gt\bay.gdb\clip_band4, image courtesy of the US Geological Survey and modified by Keranen and Kolvoord.

\EsriPress\MSDRemSen\Data\07accuracy\project1_bay_gt\bay_data_gt\int_sup, Keranen and Kolvoord.

\EsriPress\MSDRemSen\Data\07accuracy\project1_bay_gt\bay_data_gt\lc_2006_rec, Keranen and Kolvoord.

\EsriPress\MSDRemSen\Data\07accuracy\project1_bay_gt\bay_data_gt\max_sup, Keranen and Kolvoord.

\EsriPress\MSDRemSen\Data\07accuracy\project1_bay_gt\bay_data_gt\unsup, Keranen and Kolvoord.

\EsriPress\MSDRemSen\Data\07accuracy\project1_bay_gt\bay_results_gt\bay_results_gt.gdb, Keranen and Kolvoord.

\EsriPress\MSDRemSen\Data\07accuracy\project1_bay_gt\documents\07a_accuracy_bay.xlsx, Keranen and Kolvoord.

Module 7, project 2

\EsriPress\MSDRemSen\Data\07accuracy\project2_vegas_gt\documents\07b_accuracy_vegas. xlsx, Keranen and Kolvoord.

\EsriPress\MSDRemSen\Data\07accuracy\project2_vegas_gt\vegas_data_gt\landsat_ may_2006\L5039035_03520060512_MTL, image courtesy of the US Geological Survey.

\EsriPress\MSDRemSen\Data\07accuracy\project2_vegas_gt\vegas_data_gt\vegas.gdb\Vegas\ dtl.river, image courtesy of the US Geological Survey and Esri.

\EsriPress\MSDRemSen\Data\07accuracy\project2_vegas_gt\vegas_data_gt\vegas.gdb\Vegas\ dtl_wat, image courtesy of the US Geological Survey and Esri.

\EsriPress\MSDRemSen\Data\07accuracy\project2_vegas_gt\vegas_data_gt\vegas.gdb\Vegas\ mjr_hwys, from Data and Maps for ArcGIS courtesy of Esri.

MAKING
SPATIAL
DECISIONS
USING GIS
AND REMOTE
SENSING

APPENDIX C

MAKING
SPATIAL
DECISIONS
USING GIS
AND REMOTE
SENSING

APPENDIX C

\EsriPress\MSDRemSen\Data\07accuracy\project2_vegas_gt\vegas_data_gt\vegas.gdb\Vegas\ sel_sheds, image courtesy of the US Geological Survey.

\EsriPress\MSDRemSen\Data\07accuracy\project2_vegas_gt\vegas_data_gt\vegas.gdb\Vegas\ states, from Data and Maps for ArcGIS courtesy of ArcUSA, US Census, Esri.

\EsriPress\MSDRemSen\Data\07accuracy\project2_vegas_gt\vegas_data_gt\vegas.gdb\Vegas\ study_area, Keranen and Kolvoord.

\EsriPress\MSDRemSen\Data\07accuracy\project2_vegas_gt\vegas_data_gt\vegas.gdb\Vegas\ urban, courtesy of US Census Bureau.

\EsriPress\MSDRemSen\Data\07accuracy\project2_vegas_gt\vegas_data_gt\vegas.gdb\Vegas\ watersheds, image courtesy of the US Geological Survey.

\EsriPress\MSDRemSen\Data\07accuracy\project2_vegas_gt\vegas_data_gt\vegas.gdb\insup_ detrital, Keranen and Kolvoord.

\EsriPress\MSDRemSen\Data\07accuracy\project2_vegas_gt\vegas_data_gt\vegas.gdb\insup_ vegas, Keranen and Kolvoord.

\EsriPress\MSDRemSen\Data\07accuracy\project2_vegas_gt\vegas_data_gt\vegas.gdb\ LC_2006, image courtesy of the US Geological Survey.

\EsriPress\MSDRemSen\Data\07accuracy\project2_vegas_gt\vegas_data_gt\vegas.gdb\ml_ sup, Keranen and Kolvoord.

\EsriPress\MSDRemSen\Data\07accuracy\project2_vegas_gt\vegas_data_gt\vegas.gdb\rec_ lc_2006, Keranen and Kolvoord.

\EsriPress\MSDRemSen\Data\07accuracy\project2_vegas_gt\vegas_data_gt\vegas.gdb\unsup_ rec, Keranen and Kolvoord.

\EsriPress\MSDRemSen\Data\07accuracy\project2_vegas_gt\vegas_data_gt\vegas.gdb\wash_ shed, Keranen and Kolvoord.

\EsriPress\MSDRemSen\Data\07accuracy\project2_vegas_gt\vegas_data_gt\vegas.gdb\ highways.lyr, from Data and Maps for ArcGIS courtesy of Esri.

\EsriPress\MSDRemSen\Data\07accuracy\project2_vegas_gt\vegas_data_gt\vegas.gdb\ landcover.lyr, Keranen and Kolvoord.

\EsriPress\MSDRemSen\Data\07accuracy\project2_vegas_gt\vegas_results_gt\vegas_results_
gt.gdb, Keranen and Kolvoord.

Module 8, project 1

\EsriPress\MSDRemSen\Data\08urban\project1_loudoun_urban\loudoun_data_urban\bay.
gdb\bayfeatures\highways, from Data and Maps for ArcGIS courtesy of Esri.

\EsriPress\MSDRemSen\Data\08urban\project1_loudoun_urban\loudoun_data_urban\bay.
gdb\bayfeatures\loudoun, from Data and Maps for ArcGIS courtesy of ArcUSA, US Census,
Esri.

\EsriPress\MSDRemSen\Data\08urban\project1_loudoun_urban\loudoun_data_urban\bay.
gdb\bayfeatures\rivers, from Data and Maps for ArcGIS courtesy of the US Geological Survey
and Esri.

\EsriPress\MSDRemSen\Data\08urban\project1_loudoun_urban\loudoun_data_urban\bay.
gdb\bayfeatures\states, from Data and Maps for ArcGIS courtesy of ArcUSA, US Census, Esri.

\EsriPress\MSDRemSen\Data\08urban\project1_loudoun_urban\loudoun_data_urban\bay.
gdb\highways.lyr, from Data and Maps for ArcGIS courtesy of Esri.

\EsriPress\MSDRemSen\Data\08urban\project1_loudoun_urban\loudoun_results_urban\
loudoun_results_urban.gdb, Keranen and Kolvoord.

Module 8, project 2

\EsriPress\MSDRemSen\Data\08urban\project2_vegas_urban\vegas_data_urban\vegas.gdb\
Vegas\dtl.river, image courtesy of the US Geological Survey and Esri.

\EsriPress\MSDRemSen\Data\08urban\project2_vegas_urban\vegas_data_urban\vegas.gdb\
Vegas\dtl_wat, image courtesy of the US Geological Survey and Esri.

\EsriPress\MSDRemSen\Data\08urban\project2_vegas_urban\vegas_data_urban\vegas.gdb\
Vegas\mjr_hwys, from Data and Maps for ArcGIS courtesy of Esri.

\EsriPress\MSDRemSen\Data\08urban\project2_vegas_urban\vegas_data_urban\vegas.gdb\
Vegas\sel_sheds, image courtesy of the US Geological Survey.

\EsriPress\MSDRemSen\Data\08urban\project2_vegas_urban\vegas_data_urban\vegas.gdb\
Vegas\states, from Data and Maps for ArcGIS courtesy of ArcUSA, US Census, Esri.

MAKING
SPATIAL
DECISIONS
USING GIS
AND REMOTE
SENSING

APPENDIX C

MAKING
SPATIAL
DECISIONS
USING GIS
AND REMOTE
SENSING

APPENDIX C

\EsriPress\MSDRemSen\Data\08urban\project2_vegas_urban\vegas_data_urban\vegas.gdb\ Vegas\study_area, Keranen and Kolvoord.

\EsriPress\MSDRemSen\Data\08urban\project2_vegas_urban\vegas_data_urban\vegas.gdb\ Vegas\urban, courtesy of US Census Bureau.

\EsriPress\MSDRemSen\Data\08urban\project2_vegas_urban\vegas_data_urban\vegas.gdb\ Vegas\urban_SA, Keranen and Kolvoord.

\EsriPress\MSDRemSen\Data\08urban\project2_vegas_urban\vegas_data_urban\vegas.gdb\ Vegas\watersheds, image courtesy of the US Geological Survey.

\EsriPress\MSDRemSen\Data\08urban\project2_vegas_urban\vegas_data_urban\vegas.gdb\ highways.lyr, from Data and Maps for ArcGIS courtesy of Esri.

\EsriPress\MSDRemSen\Data\08urban\project2_vegas_urban\vegas_results_urban\vegas_ results_urban, Keranen and Kolvoord.

Module 9, project 1

\EsriPress\MSDRemSen\Data\09water\project1_lake_fisher\lake_data_fisher\landsat\ Aug_11_2011\L5029038_03820110808_MTL, image courtesy of the US Geological Survey.

\EsriPress\MSDRemSen\Data\09water\project1_lake_fisher\lake_data_fisher\landsat\ Aug_2_2009\L5029038_03820090802_MTL, image courtesy of the US Geological Survey.

\EsriPress\MSDRemSen\Data\09water\project1_lake_fisher\lake_data_fisher\texas.gdb\tx_ features\reservoirs, image courtesy of the US Geological Survey and Esri.

\EsriPress\MSDRemSen\Data\09water\project1_lake_fisher\lake_data_fisher\texas.gdb\tx_ features\study_area, Keranen and Kolvoord.

\EsriPress\MSDRemSen\Data\09water\project1_lake_fisher\lake_data_fisher\texas.gdb\tx_ features\texas, from Data and Maps for ArcGIS, courtesy of ArcUSA, US Census, Esri.

\EsriPress\MSDRemSen\Data\09water\project1_lake_fisher\lake_data_fisher\texas.gdb\tx_ features\tx_counties, from Data and Maps for ArcGIS, courtesy of ArcUSA, US Census, Esri.

\EsriPress\MSDRemSen\Data\09water\lake_data_fisher\lc_2006_sa, image courtesy of the US Geological Survey.

\EsriPress\MSDRemSen\Data\09water\project1_lake_fisher\lake_data_fisher\lc_2006_sa\lc_2006_sa.yr, Keranen and Kolvoord.

\EsriPress\MSDRemSen\Data\09water\project1_lake_fisher\lake_results_fisher\fisher_results.gdb, Keranen and Kolvoord.

Module 9, project 2

\EsriPress\MSDRemSen\Data\09water\project2_mn\mn_data\landsat\june_2009\L5027029_02920090601_MTL, image courtesy of the US Geological Survey.

\EsriPress\MSDRemSen\Data\09water\project2_mn\mn_data\landsat\nov_2010\L5027029_02920101111_MTL, image courtesy of the US Geological Survey.

\EsriPress\MSDRemSen\Data\09water\project2_mn\mn_data\mn.gdb\Layers\cities, courtesy of National Atlas of the United States.

\EsriPress\MSDRemSen\Data\09water\project2_mn\mn_data\mn.gdb\Layers\lakes, courtesy of the US Geological Survey and Esri.

\EsriPress\MSDRemSen\Data\09water\project2_mn\mn_data\mn.gdb\Layers\LG_study_area, Keranen and Kolvoord.

\EsriPress\MSDRemSen\Data\09water\project2_mn\mn_data\mn.gdb\Layers\minneapolis, courtesy of National Atlas of the United States.

\EsriPress\MSDRemSen\Data\09water\project2_mn\mn_data\mn.gdb\Layers\mjr_highways, from Data and Maps for ArcGIS courtesy of Esri.

\EsriPress\MSDRemSen\Data\09water\project2_mn\mn_data\mn.gdb\Layers\mn_wi, from Data and Maps for ArcGIS courtesy of ArcUSA, US Census, Esri.

\EsriPress\MSDRemSen\Data\09water\project2_mn\mn_data\mn.gdb\Layers\study_area, Keranen and Kolvoord.

\EsriPress\MSDRemSen\Data\09water\project2_mn\mn_data\mn.gdb\Layers\temperature, Keranen and Kolvoord.

\EsriPress\MSDRemSen\Data\09water\project2_mn\mn_data\mn.gdb\Layers\water, courtesy of the US Geological Survey and Esri.

MAKING
SPATIAL
DECISIONS
USING GIS
AND REMOTE
SENSING

APPENDIX C

\EsriPress\MSDRemSen\Data\09water\project2_mn\mn_data\mn.gdb\junethermal, image courtesy of the US Geological Survey.

\EsriPress\MSDRemSen\Data\09water\project2_mn\mn_data\mn.gdb\novthermal, image courtesy of the US Geological Survey.

MAKING
SPATIAL
DECISIONS
USING GIS
AND REMOTE
SENSING

APPENDIX C

\EsriPress\MSDRemSen\Data\09water\project2_mn\mn_data\mn.gdb\USA Major Highways. lyr, from Data and Maps for ArcGIS courtesy of Esri.

\EsriPress\MSDRemSen\Data\09water\project2_mn\mn_results\mn_results.gdb, Keranen and Kolvoord.

Module 10, project 1

\EsriPress\MSDRemSen\Data\10ndvi\project1_tx_ndvi\tx_data_ndvi\landsat\Aug_11_2011\ L5029038_03820110808_MTL, image courtesy of the US Geological Survey.

\EsriPress\MSDRemSen\Data\10ndvi\project1_tx_ndvi\tx_data_ndvi\landsat\Aug_2_2009\ L5029038_03820090802_MTL, image courtesy of the US Geological Survey.

\EsriPress\MSDRemSen\Data\10ndvi\project1_tx_ndvi\tx_data_ndvi\texas.gdb\tx_features\ reservoirs, image courtesy of the US Geological Survey and Esri.

\EsriPress\MSDRemSen\Data\10ndvi\project1_tx_ndvi\tx_data_ndvi\texas.gdb\tx_features\ study_area, Keranen and Kolvoord.

\EsriPress\MSDRemSen\Data\10ndvi\project1_tx_ndvi\tx_data_ndvi\texas.gdb\tx_features\ texas, from Data and Maps for ArcGIS courtesy of ArcUSA, US Census, Esri.

\EsriPress\MSDRemSen\Data\10ndvi\project1_tx_ndvi\tx_data_ndvi\texas.gdb\tx_features\ tx_counties, from Data and Maps for ArcGIS courtesy of ArcUSA, US Census, Esri.

\EsriPress\MSDRemSen\Data\10ndvi\project1_tx_ndvi\tx_data_ndvi\texas.gdb\landcover.lyr, Keranen and Kolvoord.

\EsriPress\MSDRemSen\Data\10ndvi\project1_tx_ndvi\tx_data_ndvi\lc_2006, courtesy of the US Geological Survey.

\EsriPress\MSDRemSen\Data\10ndvi\project1_tx_ndvi\tx_results_ndvi\tx_results_ndvi.gdb, Keranen and Kolvoord.

Module 10, project 2

\EsriPress\MSDRemSen\Data\10ndvi\project2_mn_ndvi\mn_data_ndvi\landsat\june_2009\
L5027029_02920090601_MTL, image courtesy of the US Geological Survey.

\EsriPress\MSDRemSen\Data\10ndvi\project2_mn_ndvi\mn_data_ndvi\landsat\nov_2010\
L5027029_02920101111_MTL, image courtesy of the US Geological Survey.

\EsriPress\MSDRemSen\Data\10ndvi\project2_mn_ndvi\mn_data_ndvi\mn.gdb\Layers\cities,
courtesy of National Atlas of the United States.

\EsriPress\MSDRemSen\Data\10ndvi\project2_mn_ndvi\mn_data_ndvi\mn.gdb\Layers\lakes,
courtesy of the US Geological Survey and Esri.

\EsriPress\MSDRemSen\Data\10ndvi\project2_mn_ndvi\mn_data_ndvi\mn.gdb\Layers\LG_
study_area, Keranen and Kolvoord.

\EsriPress\MSDRemSen\Data\10ndvi\project2_mn_ndvi\mn_data_ndvi\mn.gdb\Layers\
minneapolis, courtesy of National Atlas of the United States.

EsriPress\MSDRemSen\Data\10ndvi\project2_mn_ndvi\mn_data_ndvi\mn.gdb\Layers\mjr_
highways, from Data and Maps for ArcGIS courtesy of Esri.

EsriPress\MSDRemSen\Data\10ndvi\project2_mn_ndvi\mn_data_ndvi\mn.gdb\Layers\mn_
wi, from Data and Maps for ArcGIS courtesy of ArcUSA, US Census, Esri.

EsriPress\MSDRemSen\Data\10ndvi\project2_mn_ndvi\mn_data_ndvi\mn.gdb\Layers\water,
courtesy of the US Geological Survey and Esri.

EsriPress\MSDRemSen\Data\10ndvi\project2_mn_ndvi\mn_data_ndvi\mn.gdb\LC_2006,
courtesy of the US Geological Survey.

EsriPress\MSDRemSen\Data\10ndvi\project2_mn_ndvi\mn_data_ndvi\mn.gdb\USA   Major
Highways.lyr, from Data and Maps for ArcGIS courtesy of Esri.

EsriPress\MSDRemSen\Data\10ndvi\project2_mn_ndvi\mn_results_ndvi\mn_results_ndvd.
gdb, Keranen and Kolvoord.

MAKING
SPATIAL
DECISIONS
USING GIS
AND REMOTE
SENSING

APPENDIX C

APPENDIX D
DATA LICENSE AGREEMENT

MAKING
SPATIAL
DECISIONS
USING GIS
AND REMOTE
SENSING

Important: Read carefully before opening the sealed media package.

Environmental Systems Research Institute, Inc. (Esri), is willing to license the enclosed data and related materials to you only upon the condition that you accept all of the terms and conditions contained in this license agreement. Please read the terms and conditions carefully before opening the sealed media package. By opening the sealed media package, you are indicating your acceptance of the Esri License Agreement. If you do not agree to the terms and conditions as stated, then Esri is unwilling to license the data and related materials to you. In such event, you should return the media package with the seal unbroken and all other components to Esri.

Esri License Agreement

This is a license agreement, and not an agreement for sale, between you (Licensee) and Environmental Systems Research Institute, Inc. (Esri). This Esri License Agreement (Agreement) gives Licensee certain limited rights to use the data and related materials (Data and Related Materials). All rights not specifically granted in this Agreement are reserved to Esri and its Licensors.

Reservation of Ownership and Grant of License

Esri and its Licensors retain exclusive rights, title, and ownership to the copy of the Data and Related Materials licensed under this Agreement and, hereby, grant to Licensee a personal, non-exclusive, nontransferable, royalty-free, worldwide license to use the Data and Related Materials based on the terms and conditions of this Agreement. Licensee agrees to use reasonable effort to protect the Data and Related Materials from unauthorized use, reproduction, distribution, or publication.

Proprietary Rights and Copyright

Licensee acknowledges that the Data and Related Materials are proprietary and confidential property of Esri and its Licensors and are protected by United States copyright laws and applicable international copyright treaties and/or conventions.

Permitted Uses

Licensee may install the Data and Related Materials onto permanent storage device(s) for Licensee's own internal use.

Licensee may make only one (1) copy of the original Data and Related Materials for archival purposes during the term of this Agreement unless the right to make additional copies is granted to Licensee in writing by Esri.

Licensee may internally use the Data and Related Materials provided by Esri for the stated purpose of GIS training and education.

Uses Not Permitted

Licensee shall not sell, rent, lease, sublicense, lend, assign, time-share, or transfer, in whole or in part, or provide unlicensed Third Parties access to the Data and Related Materials or portions of the Data and Related Materials, any updates, or Licensee's rights under this Agreement.

Licensee shall not remove or obscure any copyright or trademark notices of Esri or its Licensors.

Term and Termination

The license granted to Licensee by this Agreement shall commence upon the acceptance of this Agreement and shall continue until such time that Licensee elects in writing to discontinue use of the Data or Related Materials and terminates this Agreement. The Agreement shall automatically terminate without notice if Licensee fails to comply with any provision of this Agreement. Licensee shall then return to Esri the Data and Related Materials. The parties hereby agree that all provisions that operate to protect the rights of Esri and its Licensors shall remain in force should breach occur.

Disclaimer of Warranty

The Data and Related Materials contained herein are provided "as-is," without warranty of any kind, either express or implied, including, but not limited to, the implied warranties of merchantability, fitness for a particular purpose, or noninfringement. Esri does not warrant that the Data and Related Materials will meet Licensee's needs or expectations, that the use of the Data and Related Materials will be uninterrupted, or that all nonconformities, defects, or errors can or will be corrected. Esri is not inviting reliance on the Data or Related Materials for commercial planning or analysis purposes, and Licensee should always check actual data.

MAKING
SPATIAL
DECISIONS
USING GIS
AND REMOTE
SENSING

APPENDIX D

MAKING
SPATIAL
DECISIONS
USING GIS
AND REMOTE
SENSING

APPENDIX D

Data Disclaimer

The Data used herein has been derived from actual spatial or tabular information. In some cases, Esri has manipulated and applied certain assumptions, analyses, and opinions to the Data solely for educational training purposes. Assumptions, analyses, opinions applied, and actual outcomes may vary. Again, Esri is not inviting reliance on this Data, and the Licensee should always verify actual Data and exercise their own professional judgment when interpreting any outcomes.

Limitation of Liability

Esri shall not be liable for direct, indirect, special, incidental, or consequential damages related to Licensee's use of the Data and Related Materials, even if Esri is advised of the possibility of such damage.

No Implied Waivers

No failure or delay by Esri or its Licensors in enforcing any right or remedy under this Agreement shall be construed as a waiver of any future or other exercise of such right or remedy by Esri or its Licensors.

Order for Precedence

Any conflict between the terms of this Agreement and any FAR, DFAR, purchase order, or other terms shall be resolved in favor of the terms expressed in this Agreement, subject to the government's minimum rights unless agreed otherwise.

Export Regulation

Licensee acknowledges that this Agreement and the performance thereof are subject to compliance with any and all applicable United States laws, regulations, or orders relating to the export of data thereto. Licensee agrees to comply with all laws, regulations, and orders of the United States in regard to any export of such technical data.

Severability

If any provision(s) of this Agreement shall be held to be invalid, illegal, or unenforceable by a court or other tribunal of competent jurisdiction, the validity, legality, and enforceability of the remaining provisions shall not in any way be affected or impaired thereby.

Governing Law

This Agreement, entered into in the County of San Bernardino, shall be construed and enforced in accordance with and be governed by the laws of the United States of America and the State of California without reference to conflict of laws principles. The parties hereby consent to the personal jurisdiction of the courts of this county and waive their rights to change venue.

Entire Agreement

The parties agree that this Agreement constitutes the sole and entire agreement of the parties as to the matter set forth herein and supersedes any previous agreements, understandings, and arrangements between the parties relating hereto.

MAKING
SPATIAL
DECISIONS
USING GIS
AND REMOTE
SENSING

APPENDIX D

MAKING

SPATIAL

DECISIONS

USING GIS

AND REMOTE

SENSING

APPENDIX E
INSTALLING THE DATA AND SOFTWARE

Making Spatial Decisions for GIS and Remote Sensing includes a DVD containing maps and data. A free, fully functioning 180-day trial version of ArcGIS 10.1 for Desktop Advanced software can be downloaded at http://www.esri.com/MSDRemSenforArcGIS10-1. You will find an authorization number printed on the inside back cover of this book. You will use this number when you are ready to install the software.

If you already have a licensed copy of ArcGIS 10.1 for Desktop Advanced software installed on your computer (or have access to the software through a network), do not install the trial software. Use your licensed software to do the exercises in this book. If you have an older version of ArcGIS software installed on your computer, you must uninstall it before you can install the software that is provided with this book.

.NET Framework 3.5 SP1 must be installed on your computer before you install ArcGIS 10.1 for Desktop software. Some features of ArcGIS 10.1 for Desktop software require Windows Internet Explorer version 8.0. If you do not have Internet Explorer version 8.0, you must install it before installing the ArcGIS 10.1 for Desktop software.

Installing the exercise data

Follow the steps below to install the exercise data.

1. Put the data DVD into your computer's DVD drive. A splash screen will appear.

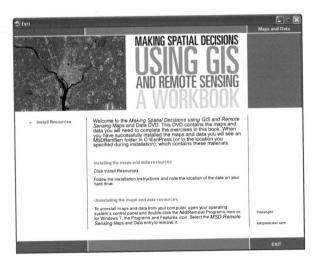

2. Read the welcome, and then click the **Install Resources** link. This launches the InstallShield Wizard.

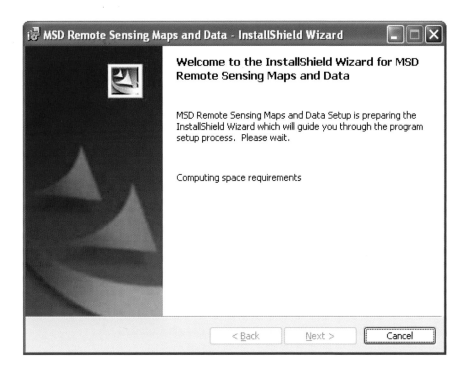

MAKING
SPATIAL
DECISIONS
USING GIS
AND REMOTE
SENSING

APPENDIX E

3. Click **Next**. Read and accept the license agreement terms, and then click **Next**.
4. Accept the default installation folder or click **Browse** and navigate to the drive or folder location where you want to install the data.

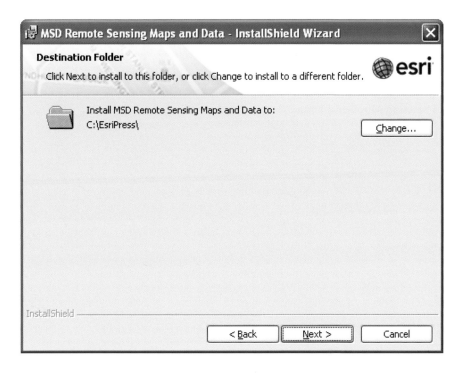

5. Click **Next**. The installation will take a few moments. When the installation is complete, you will see the following message.

MAKING
SPATIAL
DECISIONS
USING GIS
AND REMOTE
SENSING

APPENDIX E

6. Click **Finish**. The exercise data is installed on your computer in a folder called **C:\EsriPress\ MSDRemSen\Data**.

Uninstalling the data and resources

To uninstall the data and resources from your computer, open your operating system's control panel and double-click the Add/Remove Programs icon. In the Add/Remove Programs dialog box, select the following entry and follow the prompts to remove it:

MSD Remote Sensing Maps and Data

Installing the software

Note: *If you already have a licensed copy of ArcGIS 10.1 for Desktop Advanced software installed on your computer or have access to the software through a network, use it to do the exercises in this book instead of installing the 180-day trial version.*

MAKING
SPATIAL
DECISIONS
USING GIS
AND REMOTE
SENSING

To obtain the 180-day free trial of ArcGIS 10.1 for Desktop Advanced software:

1. Check the system requirements for ArcGIS to make sure your computer has the hardware and software required for the trial: http://www.esri.com/arcgis101sysreq.
2. Uninstall any previous versions of ArcGIS for Desktop that you already have on your computer. For best results, please download and use the ArcGIS Uninstall Utility from the download page.
3. Go to http://www.esri.com/MSDRemSenforArcGIS10-1, and then follow the instructions for obtaining the software.

Note: *You must have an Esri Global Account to receive your free trial software. Click* **Create an account** *if you do not have one. When prompted, enter your 12-character authorization number, which can be found printed on the inside back cover of this book.*

Assistance, FAQs, and support for your trial software are available on the online resources page at http://www.esri.com/trialhelp.

Uninstalling the software

To uninstall the software from your computer, use Add/Remove Programs from your operating system's control panel. Select the following entry and follow the prompts to remove it:

ArcGIS 10.1 for Desktop